SPACE, TIME AND THE ETHICAL FOUNDATIONS

"Robert Allinson's book represents tremendous thoughtfulness, originality, and erudition. Its wide-ranging and lucid discussions cover a huge terrain, from ancient metaphysics to quantum mechanics. The enlistment of certain classical Confucian concepts and themes at critical junctures to advance the book's argument also provides luminous comparison. His interpretation of the Confucian emphasis on life as social and self-preservation is both humane and interesting, much as his analysis of the Mencian notion of compassion deserves our attention."

Anthony C. Yu, Carl Buck Distinguished Professor in Humanities, Chairman, Division of East Asian Languages, University of Chicago, Divinity School.

In Space, Time and the Ethical Foundations ideas about space and time are developed, unique to the history of philosophy, that match the new physics. A well grounded metaphysics is presented which offers a safe haven between stifling scepticism and wild imagination, and an original philosophical method is demonstrated which sharply demarcates philosophy from the empirical sciences.

A new foundation is laid for ethics by grounding ethics on the author's psycho-biological deduction of the emotions that offers a progressive model to replace the Freudian paradigm. An originally designed trans-cultural ethics, doubly grounded on both Eastern and Western thought, presents an antidote to the contemporary retreat into relativism. Insights from biology, psychology, evolutionary theory and ethics are brought together in a unique and fruitful synthesis. At the same time, human barbarisms such as the Holocaust are pointed to as reminders that there are just limits to compassion. This book presents a sophisticated text for metaphysics, epistemology and systematic ethics.

Professor Robert E. Allinson is the author or editor of many books including, *A Metaphysics for the Future*, 2001 and *Understanding the Chinese Mind: The Philosophical Roots*, New York: Oxford University Press, 2000 (Tenth Impression). He serves on eight editorial boards including *The Journal of Chinese Philosophy and Asian Philosophy*. His work has been translated into Chinese, Japanese, French, German and Italian. Professor Allinson has been a Senior Associate Member of St. Antony's College, Oxford University, an Associate Member of Balliol College of Oxford University and Visiting Fellow at the Graduate School of the Department of Philosophy at Yale University. He is a Fellow of Shaw College and a professor in the Department of Philosophy at The Chinese University of Hong Kong.

For Irènè, whose constant support made all this possible.

Space, Time and the Ethical Foundations

ROBERT ELLIOTT ALLINSON
Department of Philosophy
The Chinese University of Hong Kong

Routledge
Taylor & Francis Group
LONDON AND NEW YORK

First published 2002 by Ashgate Publishing

Reissued 2018 by Routledge
2 Park Square, Milton Park, Abingdon, Oxon OX14 4RN
711 Third Avenue, New York, NY 10017, USA

Routledge is an imprint of the Taylor & Francis Group, an informa business

A Library of Congress record exists under LC control number: 2001099155

ISBN 13: 978-1-138-74073-0 (hbk)
ISBN 13: 978-1-315-18330-5 (ebk)

In grateful acknowledgement of the receipt of the Vice-Chancellor's Special Fund for Excellence granted by The Vice-Chancellor of The Chinese University of Hong Kong which was a key source of encouragement and support of the writing of this book; for the graciousness of the esteemed professors, Anthony C. Yu, Carl Buck Distinguished Professor in Humanities and Chairman, Division of East Asian Languages at the University of Chicago, John E. Smith, Clark Emeritus Professor of Philosophy at Yale University and Lewis Hahn, Editor of that illustrious series, *The Library of Living Philosophers*, who were kind enough to read earlier versions of this work; and in appreciation for my wife, Irènè, whose faith in me has been the mainstay of this project from its beginning to its completion.

'In a certain sense, therefore, I hold it true that pure thought can grasp reality, as the ancients dreamed.'
Albert Einstein, *The World As I see It*, New York: Covici Friede Publishers, 1934, pp. 36-7.

'It is not thy duty to finish the work. But thou are not at liberty to neglect it.'
Pirke Abot (*The Wisdom of the Fathers*) [one of the treatises of the *Talmud*], Judah Goldin (trans.), New York: The New American Library, 1957, p. 116.

'*Et quam multa queas, fac quoque multa velis.*'
If you can do much, you must will much.
Antonio Ludovico Muratori, 1743.

O somma luce, che tanto ti levi
da' concetti mortali, alla mia mente
represta un poco di quel che parevi

e fa lingua mia tanto possente
ch' una favilla sol della tua gloria
possa lasciare alla futura gente...

O light supreme, who so far dost uplift
thee over mortal thoughts, lend again to
my mind a little of what thou didst seem

and give my tongue such power that it may
leave a single sparkle of thy glory to
future men...
Dante, the last canto of the *Divina Commedia.*

Contents

PART I

In Between Metaphysics and Epistemology

Largo espressivo
With grave dignity

1 On Certifying What is True in Philosophical Discourse

There is a certain advantage in outlining one's methodology for establishing truth and one's criteria of truth at the beginning of one's investigation. There must be a prior agreement, whether implicit or explicit, as to how truths are to be recognized, accepted, or otherwise known as truths before there can be broad agreement among followers of a discipline as to what will be considered the accepted truths in or of that discipline. There is no value in making truth claims in any discipline in the absence of any methodology for establishing agreement on how what is to be taken as true is known to be true.

In the event that a new, philosophical claim does not fit either pre-established, co-temporary standards for truth, or does not make a case for the establishment or the acceptability of new standards, any claim that is lodged will come to be taken, quite rightly, as an assertion. Why should one assertion hold any claim to allegiance as over against its opposite? Therefore, it would be advisable to explore what means can be employed to certify what is claimed to be true as being true, prior to the lodging of any substantive claims.

It may be alleged that even to propose to begin an inquiry with an outline of a truth finding and truth certifying methodology is to begin with an assumption. Is it not being assumed that there is some means by which a veridical truth claim can be established? While this may appear to be an assumption, it is to be hoped that in what is to follow that this will not stay at the level of an assumption, but will clearly become a part of what is certified. If one were to treat what has been said so far as part of inevitable assumption making, there would be no point at all in beginning from the starting point of searching for the means whereby truth may be certified. One may as well initiate one's inquiry by asserting the rawest of uncertifiable metaphysical claims, or simply abide by the truths or the criteria for acceptable truths that are already unquestioned and unquestionable in the academy. It is therefore to be hoped that what may appear as an assumption; e.g., that truth can be recognized or that there is truth, will show itself not to be an assumption, but will prove itself to be an undeniable and inescapable fact. The proposal which is being advanced here is that, with due qualification, what is true can be discovered to be true. But, this is no mere assumption. It is a phenomenological fact. How it is a fact rather than an assumption is something that will gradually emerge in the sequel. While the above statement, 'what is true can be discovered to be true' might appear to be a trivial truth or a tautology, the point of the statement is that truth is discoverable and is not a product of an inference which in turn is based on a number of unproven and unprovable assumptions. That there is something which can be referred to by the word 'truth' is itself a most unpopular standpoint today. *'Truth' is a taboo.* To be in line with current fashion one must speak of cultural beliefs or at best heuristic and pragmatic belief systems rather than to commit oneself to the pursuit of that elusive quarry of truth. To go even further and propose that not only do truths exist but that they are discoverable and that one can outline a methodology of truth discovery is to commit oneself to a most grave form of political incorrectness.

The proposal is that it is possible to discover a truth or set of truths, which will be revelatory of all human beings as such upon introspection of one's own subjective insights. This appears to be a kind of assumption which has something of the form: whatever the subject knower discovers to be true about herself or himself has a universalizable applicability. But in reality, this proposal is not an assumption like this at all. In order to discover, however, that this proposal is not an assumption, a putative set of truths which are being held to be subjective universal truths will have to be set out for the purpose of putting this proposal to the test. Only after testing such putative (at this stage) subjective, universal truths will a position will be developed to ascertain that no assumptions whatsoever are being made in the process of staking claims as to what constitutes truth.

After performance of the *epoché*, whatever one discovers by personal introspection to be an apodeictic, phenomenological truth (within the parameters of one's consciousness), is capable of so being discovered by any and all humanity. How does one 'by personal introspection' become aware of a phenomenological truth that is apodeictic? Once the phenomenological standpoint has been obtained, a truth that discloses itself in the form of 'that which can never be otherwise' is an apodeictic, self-evidencing truth. It is not the personal introspection that is the source of the truth: the personal introspection is simply the means by which one may obtain access to the quality of 'can never be otherwiseness'. When a truth is cognized as inescapably true and for which no counter-example can be envisioned, it is that type of truth which may be considered a necessary truth of consciousness.

The statement that began the previous paragraph possesses the appearance of an assumption, but it is not being put forward as an assumption. It also appears that another assumption has been made, that human beings are truth-discovering animals. This, however, is also not an assumption despite its appearance of being one. How is either of these claims known to be true? Phenomenological introspection on the part of each and any philosophical reader is the only test for knowing whether either of the foregoing truth claims is true. Without this collaboration on the part of the subject reader, any and all claims, which are set forth within this work, will appear to be ungrounded assumptions.

It is a human capacity to know that whenever one is able to discover some necessary truth about oneself, that selfsame truth is at the same time reflective of and vouchsafed to be true of the human condition, *per se*. This is the philosophical counterpart to Adam and Eve's charge and dispensation to name the creatures of nature in *Genesis*. This is not an assumption, since it can be immediately tested in the consciousness of each subject reader. To claim that each subject knower is a truth-discovering and truth-certifying animal appears as the rawest of assumptions. Is it not an assumption to state that each subject knower possesses a capacity for recognizing what is true in each truth claim within philosophical discourse? No, for what in this case appears to be an assumption or an assertion which requires objective confirmation is, on the contrary, a description of human experience in the private experience of one subject knower, which is a revelation about human nature that is at once accessible to every subject knower. Any human being, speaking from the depths of experience, is speaking as all human beings, since human nature is universal. Man is a truth-bearing and truth-validating animal. Man is, in short, an epistemological animal. Any necessary, subjective truth is at once a reflection of the universal human condition. The only real issue is first acknowledging the universal nature of man. And this is not as difficult to do as it first appears. It only requires setting aside for a moment a most widespread and deep-seated intellectual prejudice that there is no such thing as a universal nature of man.

Suppose one commences with the familiar, very elementary truth of arithmetic, that, 1 + 1 = 2. It is an undeniable fact that each human being can and in fact must make this discovery within herself or himself.[1] It can quickly be countered that such a statement is based on every party to this truth accepting the meaning of the symbols used. This, too, is an undeniable fact. (In fact, this objection itself becomes another illustration of a necessary truth). But the more significant point is that once the meaning of the symbols has been established and agreed upon, all parties possess the capacity for recognizing, and in fact of being compelled to recognize the claim as being true. One may attempt to derogate the importance of this discovery by referring to this as an analytic, or empty truth, but the more important point relevant to the argument being advanced herein is that the possibility of such a recognition and certification at once demonstrates a universal capacity for recognizing and thus certifying the truth claim that is lodged, regardless of what kind of truth it is alleged to be. In a word, there is a universal truth-knowing ability.

Now, while one could argue endlessly whether such a truth as a mathematical truth is an empty analytic truth or a synthetic truth as Kant alleges, there is no need, at the present moment, to enter into such a debate.[2] For present purposes, all that is necessary is to recognize that whatever type of truth one takes a simple mathematical truth to be; it is a universally certifiable truth claim. This, in and of itself, is a remarkable fact.

What is being proposed in this work is that there are a set of truths, with more substance or complexity than the truth that 1 + 1 = 2, that can also be recognized as being universally true. In the case of the simple truth of mathematics, where nothing much is at stake, there is no questioning this possibility. Some may counter that even these so-called "simple" truths such as 1 + 1 = 2 are debatable and are dependent upon the choice of certain systems of mathematics. Bertrand Russell, co-author of *Principia Mathematica*, takes exception to this in his *The Problems of Philosophy*.[3] The point being made in this present, respective work is that whatever the system is within which 1 + 1 = 2, or by which 1 + 1 = 2 is derived, this system also must be a system in which its own assumptions must be corroborated by an immediate intuitive act on the part of a subject knower without the recourse to exterior assumptions. That this system happens to be the system within which or by means of which a direct recognition of the truth claim being made becomes possible is not a proof that this is the best system or the only system of mathematics; it is only an empirical circumstance that this particular system happens to coincide with the immediate truth-discovering capacities of the human being. What is of interest is not the vindication of any certain mathematical system, or any particular type of truth vindication to be found within mathematical systems (logicist, formalist, intuitionist), but only the discovery that there are truths that human beings are in a position to recognize to be truths, which, for human beings, also turn out, upon phenomenological introspection, to be universal truths. In fact, it is only because they are universal truths that it is possible that they are discernible as phenomenological truths in the first place.

It is difficult to find a significant number of fellow inquirers who would agree that there are such universal, mathematical truths. It is even more difficult to find significant agreement that such universal mathematical truths illustrate a universal capacity for truth discovery within mathematics that is shared by all human and perhaps all rational beings. The greatest difficulty that remains is to find fellow inquirers who are willing to agree that there are any universal statements that can be made about the human condition as such. Once it is granted, however, that human beings do possess a truth-discovering ability, no matter how trivial this truth-telling ability is considered by those who grudgingly agree to its existence, one is thereby committed to the reality that human beings possess a truth-discovering and truth-certifying capacity, and this is no small advance. What has been established is that truth, contained in the statement of one human subject, possesses a universalizable applicability. That this is a qualified truth, e.g., that one must first recognize and know how to interpret certain symbols, or subscribe to a certain mathematical system, is of no more moment than that one must first share and therefore know how to interpret the signs of a common language before one can communicate or present a truth. The capacity for language sharing is not identical to the capacity for truth discovery. Once it is agreed upon that human beings do possess a universal truth-discovering capacity, it is no longer a mere assumption that human beings are truth-recognizing and truth-validating animals. It is already an established truth. What may at first have appeared as a "wild" metaphysical claim is in fact a well-founded phenomenon.

The first truth, that 1 + 1 = 2, may be argued to be either analytic or synthetic. But, what cannot be denied is that it is recognizably a truth of some sort. The second truth, that human beings therefore possess a truth-knowing capacity is not a statement that can be reduced to an analytic truth. It is a truth about human beings, which is discovered by and in experience. It is a not a truth that is discovered via the analysis of the meaning of language. And it is an undeniable truth. This second truth is not reducible to knowing the meaning of the symbols that are in use. It is a truth about human nature as such. Thus, the promise made at the outset of this chapter that the truth which was claimed to be a truth would be shown not to be a mere assumption, namely, that all human beings as such possess a truth-discovering and truth-certifying potential, has been fulfilled.

Here, this truth-discovering ability has been labeled a human capacity. The discovery that human beings possess this truth-telling ability has been labeled a phenomenological discovery. What is meant by such a high-sounding phrase? What is meant is that it can be detected, from a simple subjective test, and only from a simple subjective test, that what each subject inquirer understands as necessarily true, is true of the human condition as such. There is no further proof of this, and there needs to be no such proof. Each subject inquirer who recognizes that all human beings possess a truth-discovering capacity is making a phenomenological discovery, which is at the same time a universal truth. *There is no other means of access to the understanding of this truth as a necessary truth which is understood as true and understood as necessarily true in the same epistemic moment. One is recognizing that a subjective experience of what is seen to be true by the subject knower can, nay, indeed must be seen to be true by all subject knowers.*

It is no mean feat that philosophers can now say that they have at once discovered something that is true, that is not merely a truth within mathematics (that human beings possess a truth discovering capacity), and that this discovery also reveals that philosophers at the same time have discovered a universal truth. What is more, a philosophical methodology has been established in the process. One now knows that if one were able to establish what is true, that a method for establishing what is true can also be articulated. Certain kinds of truths, which are understood to be true subjectively, have a universalizable applicability. In order to be able to know a "subjective" truth to be a universal truth, it must be the case that it is not known to be true in virtue of its being known subjectively, but in virtue of its subjective discovery existing as a universally shared capacity. Its subjective discovery conditions are a proof of subjective universality. This is an example of the beginnings of a philosophical method for the articulation of how one can establish and recognize philosophical truths.

Needless to say, when the realm of more volatile truth "claims" has been entered, agreement will not be so easily reached. What is interesting, however, is that some headway has been made. This amount of headway is already sufficient to dislodge the current claim that all that exists is subjective opinion. This point has already been gainsaid.

All human beings possess universal truth-knowing powers, however restricted one may wish to allege such truth-knowing powers to be. And this very acknowledgement, that human beings possess truth-knowing powers, is a very powerful admission. For it grants that what is subjectively seen as true can have a universalizable applicability. One human being can set forth and recognize a truth that must be agreed upon by any and every human being. And this is no trivial discovery. It is a discovery that human nature, while subjective in each subject knower, is also at the same time universal. It follows that truth must be discovered subjectively, but truth is not subjective. Of course, all of this depends upon the recognition of the simple fact that all human beings can and in fact must recognize that $1 + 1 = 2$. That other mathematical systems can be devised in which this simple truth is no longer a truth does not affect the truth of $1 + 1 = 2$, which is a phenomenological truth. Of course, it can be argued that it may take some training before one can recognize that $1 + 1 = 2$. However, with the minimum of such training, such a recognition is unavoidable. It could be argued that this is but a simple truth. The simplicity of a truth does not render it any the less profound. However, the weight of intellectual prejudice is often difficult to remove.

Inadvertently, philosophers have discovered the universalizable quality of human nature in the midst of discovering one universal truth. Philosophers may have discovered something else as well, that the method of going from the subjective to the universal, rather than being an antiquated superstition, is a method that needs to be revived, dusted off and reinspected. For, it has now been discovered that human nature is universal such that the subjective, in such cases as qualified above, is a valid clue to the objective.

A method for finding truth has been discovered in the process of starting with subjective truth conditions. It has been discovered that the way in which one makes and establishes universalizable claims is through subjective introspection and subjective insight. No one can communicate to a potential subject knower that $1 + 1 = 2$. This is a discovery which must be made by each subject knower. One can, of course, simply accept the mathematical equation as true. But, this is not the same as seeing it to be necessarily true. If it were to be seen as true, it would also thereby be seen that every subject knower must also see it as true. What can be seen as necessarily true, subjectively, possesses a universal status. The foregoing provides a great insight into philosophical method. Once a mathematical or a philosophical truth has been discovered, one does not require recourse to further outside proof.

This is not a truth from mathematics. It is a truth about how one finds out what is true in mathematics.

One finds out what is true in mathematics (at least in the example cited) by subjective discovery. One is not in need of outside verification. In fact, outside verification is of no utility. It is of no assistance in the discovering of this truth (or even of the first truth that 1 + 1 = 2). The truth of 1 + 1 = 2, and the sequence of its reflective correlates - that this is a universal truth, that one has thereby also discovered how such a truth is known, that through becoming aware of this one has thereby also discovered that human beings are truth finders and truth knowers - have one and all been discovered and simultaneously established within consciousness - in each subject knower's conscious recognition or insight. That all of these truths - or even the first truth taken by itself - can be corroborated by every human subject is simply a proof that subjective insight can be universally valid. Corroboration is not a proof of the validity of subjective insight. It is not a confirmation. Each corroboration is an additional or correlative discovery that possesses no confirmation value.

When one sees that two straight lines can never enclose a two-dimensional space, one sees this in the form of 'once true, always true'. There is no need to test this truth over and over again. Its truth-value will not be enhanced through repetitive testing. No empirical truth is known in the form of 'once true, always true'. The very moment of knowing that two straight lines can never enclose a two-dimensional space is at once a knowing of this truth and a knowing of this truth as a necessary truth. Its truth-value is epistemically inseparable from its truth modality. It may be claimed that it is possible to become aware of the truth of the proposition that 'two straight lines can never enclose a space' without becoming aware of the necessity of this truth, but if one carefully analyzes the proposed experiential case, one will find that one is not fully cognizant of the truth value of the proposition without being aware of its necessity and consequently one is not truly aware of its truth without knowing its modal value.

The real verification of philosophical truth lies in the subjective discovery that is at once a discovery and a validation of the truth in question. Such discoveries can be rediscovered; in fact, they must be capable of being rediscovered if they are to lay claim to truth status. But, such rediscovery is not validation. It is not confirmation. It is a fresh discovery in every instance. At the same time, it is a further illustration, or an additional enactment of what has already been found to be true. It may be taken to be corroboration, but this is accidental to its essence. Its essence is a further, subjective re-enactment of the truth.

And, the truth is known and only known in each individual, subjective enactment. It is not known in any other or from any other subjective re-enactment.

What is most telling about this discovery about the nature of philosophical truth is that this same test applies to the higher truths about human nature that the mathematical truth reveals, e.g., all human beings possess truth-knowing capacities. And *this* is also a truth, which is not *known* to be true in or by the verification of it by others. Its discovery by others is simply another appearance of the selfsame truth. It is known to be true in and by each subject knower in the privacy and originality of her or his own consciousness.[4] Others must confirm this truth for themselves in their own subjective re-enactments.

And so it has not merely been discovered that human beings are truth-discovering animals as far as mathematical truths are concerned. It has at once been discovered that human beings are truth-discovering animals *per se*. This level of discovery is a metaphysical truth. That the human animal is a truth-discovering animal is not a truth that can be discovered or proven by any empirical means. Empirical means can only demonstrate that this is a capacity that belongs to human beings circumstantially. Empirical means can never demonstrate any existential truth to exist by necessity.

In fact, since the primordial discovery of truth knowing is verifiable only in the act of truth discovery, the fact that human beings are truth-certifying animals can and must be proven by pure philosophical means alone. Even the best of empirical tests always relies upon and can only be explained by reference to a prior understanding of the original act of truth knowing which is possible only in an original act of discovery/validation.

By reflecting upon the conditions of mathematical truth knowing, one can and must discover at the same time a truth about human nature itself. A metaphysical truth, that the human being is a truth-knowing animal, is discovered and established at once. It is not empirically provable, for every confirmation is not a confirmation as such. It is an illustration of the selfsame truth. That it is an illustration and not a confirmation is a part and parcel of the fact that it is a universal truth. A confirmation, *simpliciter*, would imply that it was not known to be a truth in the first place, but was simply a theoretical speculation or a tentative, empirical hypothesis, which had been waiting upon further, empirical corroboration for its validation. In this case, what was set forth at the outset was that which was already known to be true. Each empirical "confirmation" is simply another illustration of the same truth, not a further confirmation.

Metaphysical truths are self-revelatory. They are self-certifying. When any subject knower stumbles upon the same truth-discovering possibility as the initial discoverer, that subject knower is not confirming, validating or verifying the truth of the first discoverer. Other subject knowers are re-enacting the same subjective discovery. When the discovery is made for the first time - and for every subject there must be a first time - (to discover, not to learn to repeat from memory, that, for example, 1 + 1 = 2), at the same time a universal truth about human nature as such can be discovered, which can also attend the initial discovery. One is not in need of other human beings to make the same discovery. *Truth is never social; truth is a private affair.*

In fact, that other human beings can make the same discovery, it is necessary for those other human beings to satisfy the truth conditions set forth in this work. But, if other human beings could make this discovery, it would provide no assistance or assurance to the first discoverer that her or his discovery was valid. That other human beings can and in fact *must* make the same discovery is not the proof of the truth of the first discovery. Each subject knower must make her or his discovery for herself or himself.

In the meantime, has anything else been discovered? It seems that the discoveries made so far include the fact that human beings one and all possess a truth-discovering capacity, and, what is apparently even more controversial nowadays, that human nature is a universal. It is not fashionable to speak about the universality of human nature. The notion of a human "nature" is much maligned. It is presumably a metaphysical assumption. In the foregoing, there has been some attempt to demonstrate why this is no mere assumption. It is a condition for the very possibility of the discovery of the truth of such a simple arithmetical proposition that 1 + 1 = 2.

Philosophical methodology is a unique methodology. It is not the same as methodology in the empirical sciences. In order to discover that human beings possess a universal human nature, it is of no assistance to send out a questionnaire. It must be discovered subjectively in an original act of consciousness and then subjectively realized that within consciousness a capacity for recognizing universal truths is possessed universally.

If one were not at first to have *known* that 1 + 1 = 2, one would not be permitted, in good faith, to set this out as a claim for others to discover. One must at first recognize this as a truth of and for the individual subject knower. In making the discovery of this truth, the subject knower simultaneously happens upon another truth about human nature: that it is universal. This too, does not wait upon the testimony of others for its validation.[5] Every subject knower must make the same discovery to know it to be true. But, this only means that empirically, while someone must be the first discoverer, what inevitably makes the discovery true is its universalizable quality. That is why it is perceived as true. Otherwise, how could it have been known to be true in the first place? *It is subjectively discovered, but it is not a subjective truth.* Its truth-bearing qualities issue from its status as a universal truth. What is most remarkable is that in the very conditions of subjectivity, and in fact, only in the conditions of subjectivity, can one discover universal truths. This is another illustration of the fact that human nature is universal. This is a fact that would be difficult if not impossible to prove on its own, but incontrovertible when discovered as a condition for subjective truth discovery.

One could not initially have known an apodeictic truth unless it already were a universal truth. That it is a universal truth is already a condition for its having been apprehended as an apodeictic, subjective truth. This cannot, in and of itself be 'proved', but there is no other way of accounting for how it is possible that mere subjective conditions can be the occasion of a universalizable truth-knowing situation in the first place. That it is possible in the first place and in fact necessary, that one can, on an individual basis, and *only* on an individual basis, discover and subjectively verify a universal truth condition, must entail that the truth condition is a universal condition. Otherwise, truth conditions would not and could not be distributed on an individually subjective basis. The discovery about this simple truth in mathematics and its consequences for further truths about human nature and its truth-knowing potentialities, also reflects that such truths that are discovered have the qualities and must have the qualities of being subjective universal truth conditions.

Human nature, although distributed individually, is per force universal. The first major philosophical discovery, apart from the discovery of the mathematical truth-knowing capacity and even apart from the discovery that human nature possesses a truth-knowing capacity, is that human nature is and must be subjective, and yet at the same time is universal, and *all of this follows from the simple discovery of a mathematical truth*. Subjectivity, at least in certain specialized instances, is an index of universality. Universality, in turn, is a condition of there being certain forms of subjective truth in the first place. Otherwise, the phrase, 'apodeictic subjective truth', would be meaningless. Metaphysical truth can be discovered as a condition for the very possibility of there being apodeictic, subjective truth discovery situations in the first place.

It may be rewarding to take another look at the concept of the certification of truth in philosophy with which this chapter commenced. It may have sounded absurdly simple at the outset to have ventured the claim that arriving at philosophical truths bore a distinct resemblance (if not an identity) to knowing that 1 + 1 = 2. The arithmetic example was intended to perform two functions. First of all, it was intended to inextricably link the character of philosophical truth discovery to a subjective awareness within consciousness, rather than to an empirical observation of the external world. It was not an accident that Plato placed mathematics and philosophy above the divided line in the realm of knowledge. Secondly, it was intended to show that this subjective condition of truth discovery is universal, and hence philosophical truth claims did not require confirmation in the sense that scientific truth claims do, but were known as true in the very act of their being discovered. The moment of discovery and validation were one moment. That this is not therefore an arbitrary "intuition" which may differ from subject to subject is what can be tested. Such a test is not strictly speaking a confirmation of the truth, but an outline of the methodological procedure for the testing of the truth. What is a truth is known as true in the moment of its discovery. Every fresh and further discovery of it does not further confirm it. Every further discovery merely provides an additional illustration.

In this sense, the truths of philosophy are very much like the truths of mathematics, if one recalls the spider analogy of Bacon's. However, philosophical truths are not simply like the truths of mathematics, because they are not concerned only with the relationship among numbers (arithmetic) or the spatial configurations of objects (geometry). They are unlike mathematics, because they relate to other epistemological experiences as well. For example, in the discovery that 1 + 1 = 2, one discovers willy-nilly that human beings possess a truth-discovering capacity. This is not a truth about mathematics; it is a truth about how humans are able to discover truth within mathematics. In knowing a philosophical truth one knows not only the content of the truth, but one also knows that the truth is known as a necessary truth and one knows how one comes to know the content of the truth as a necessary truth.

But, if one finds that the reflective correlates of the initial mathematical truth are not compelling, the simple arithmetic example will serve as an example of necessity and universality. The apodeicticity that obtains is not a mere psychological feeling of certainty. *It is apodeictic, because its truth is not about, nor does it arise from, nor can it be reduced to merely this individual empirical embodiment of the truth.* The apodeicticity of the discovered truth is a quality, which cannot be justified by its empirical holder and discoverer, and thereby cannot arise from nor can it merely belong to the individual truth discoverer.

From whence can the certainty derive? One could attempt to say that the certainty derived from the simple fact of the inspection of the equation on the board or in one's imagination or on one's computer screen. But, what is known is not a relationship that exists simply in this one instantiation, but a universal relationship between all meaningful symbols of like kind. One does not simply know this particular relationship; in fact, the cognitive and apodeictic component of truth is the universal relationship that this particular illustration simply symbolizes or adumbrates. One knows more than one could possibly know from this one example alone and that is whence the certainty derives.[6] It is not simply a psychological truth (whatever this bogeyword expression means); it is a realization (even if tacit) that what is known to be true, in a truth of this kind, is always and universally true. This is the moment in which the discovery of truth and its validation are one.

The foregoing account of truth may be branded an "intuitive" account, but this act of labelling is uninformative. This may also be branded 'psychologism', but this is inaccurate since it does not distinguish this kind of truth from empirically conditioned, psychological truths, such as the alleged truth that all human beings are selfish. Most importantly, it is not true because it is felt to be true; it is true because it is realized to be true in the exact moment of understanding the mathematical relationship. *There is no difference between understanding it in the first place, and knowing it to be true.* There is no separation in knowledge in which something is known and something is known as true. The very act of knowing it is a knowing it to be true. *It cannot be known as true unless it is true, and it cannot be known to be true unless it is known.* In mathematics and in philosophy, truth discovery and truth validation are two aspects of the same phenomenon.

In order to meet the tests that have been set out for the certification of philosophical truths, every first-order truth claim that is set out must be of this kind. What is remarkable about philosophical truth claims is their universal quality. From a particular example, one appears capable of inferring a universal relationship. But, there is nothing shameful about this. It is simply a straightforward description of the methodology of philosophical truth discovery.

Why, then, if it were so simple, is there so much disagreement among philosophers as to what constitutes the truth about anything?

This, it may well be, owes at least some of its origin to the lack of clarity, up until now, regarding the procedural methodology of establishing philosophical truth. There has been a dim awareness of the foregoing exposition in that certain philosophers are acknowledged as "Great" even if one does not agree with their truth claims. This would seem to be odd unless philosophy were something like art or music where Leonardo or Beethoven are admired as great while neither make any claims to truth. But, if philosophy is concerned with truth, how can Kant be admired, if Kant is, in one's opinion, in error about much of what he claims as truth? The probable explanation for this is that, however much Kant may appear to be in error, in some respect, he has touched the truth-giving dimension of philosophy, and it is because of this that due recognition is given him. It is not simply that he has constructed a huge architectonic which can be admired for its logical intricacy and comprehensiveness, although it is a false description. It is, more accurately, because he has discovered the arena of philosophical truth discovery even if he may have been mistaken concerning the identity or the roles of the players within that arena.

It may be objected that in the lengthy explanation of the simple example from arithmetic that has been given that a very trivial observation has been elevated to the level of a profound philosophical truth. But, it is not at all clear that this is an objection. The essence of the discussion of the discovery is the importance of the discovery which, up until now, may have been languishing as a "trivial" truth.

Why is it that to know that 1 + 1 = 2 is but a trivial accomplishment? It is no trivial accomplishment to discover that which is true. It is the very heart of philosophy itself as a discipline which makes it different from music or art. *Philosophy is the art of truth discovery.* One must also recall that in the act of knowing this simple truth of arithmetic one knows not only this truth but that it is necessarily true and thus that one can know a necessary truth.[7] From the earliest days of philosophy, philosophers have attempted to demarcate the province of the human being. For example, one can recall Aristotle's definition of man as the rational animal. But, other animals reason and seem to possess symbol-making and symbol-recognizing capacities, as has been learned from the recent experiments with teaching gorillas and chimpanzees sign language and the discovery that dolphins already possess and utilize their own system of linguistic communication. These animals may even possess a truth-discovering capacity as well. And, while other animals may also possess the following attribute, man is hereby demarcated as the animal who can become aware that she or he is capable of discovering and knowing the truth, and knowing that discovering and knowing the truth are one and the same thing. In other words, man is the philosophical animal. That dolphins (or other animals) can use language is discovered empirically by human beings and this discovery remains for the human observer an empirical truth. To understand that dolphins can apply the concept of truth would itself require a dolphin-internal metaphysical recognition of truth, which cannot be the object of an empirical discovery. Whether dolphins can make any of these discoveries about themselves is a possibility which may and must be left to the dolphins to discover or to comment upon.

Notes

1. To avoid the problem that all mathematical truths and the consequent reflection upon their epistemic status could nonetheless be errors forced upon one by a *malin génie* requires for Descartes an initial belief in the goodness of the Deity. While it is most unfashionable to base mathematical certitude upon the premise of a good Deity who is more powerful than any evil genius, it is unclear how without this hypothesis one can escape from the omnipossibility of systematic deception. While it could always remain unknown which truths were really deceptions, without the existence of a good Deity it would never be known that the truths of mathematics (and reflections upon them) were reliable. This question will be reserved for a later discussion in the sequel.
2. For Frege, the laws of arithmetic are most probably analytic *a priori*. However, he sees no conflict between holding this view and holding the view that they are nonetheless laws of the laws of nature. *Cf.*, Gottlob Frege, *The Foundations of Arithmetic*, J. L. Austin (trans.), Evanston: Northwestern University Press, 1980, p. 99. That the laws of at least some mathematics apply to the physical world would seem to argue against their being merely or exclusively analytic or definitional truths. At the very least, it would appear that their application to the world would then be a remarkable coincidence. It could be argued that mathematical truths were known (or could be known) analytically, but that they were not analytic truths *simpliciter* in terms of being true only because of conventional semantics. It could be argued that the fact that human beings could know mathematics analytically revealed that analytic knowledge was analytic precisely because it was definitional that human beings, as intellectual microcosms and micro-intellectual reflectors and repositories, stored and knew analytically, the laws of the universe. For the human mind, what was analytic was at the same time synthetic in the sense of also being true of the universe. But, analytically grasped truths were the human mind's mode of knowing that which was true of the universe as a whole. This might serve as an answer to Sir Roger Penrose's question of '... why do physical laws seem so accurately to follow ... precise and subtle mathematical

descriptions?' *Cf.*, Roger Penrose, *Shadows of the Mind*, New York: Oxford University Press, 1994, p. 208. The 'mental laws' of the human mind are, at least in some instances, intellectual reflections of the physical laws of the universe. Recently, Michael Perry of the University of Bristol has argued that there is evidence that one of the last holdouts of mathematics, prime numbers, now also are found to be mirrored in the laws of quantum physics. Moreover, laws of quantum physics are now being employed to discover more about the nature of prime numbers. The expressions 'analytic truths' and 'synthetic truths' are sometimes employed instead of 'analytic judgements' and 'synthetic judgements' both for the sake of simplicity and for the sake of greater accuracy. From the point of view of the present work, these truths are not true in virtue of the kind of judgement rendered, but are true in virtue of how their truth is contained in the *cognoscendum*. The judgement is made on the basis of what kind of truth is ascertained. The truth is not a result of the judgement. Moreover, if one leaves the description of 'synthetic' or 'analytic' at the level of judgement or proposition, one is still one step removed from the truth-ascertaining moment since a judgement or proposition must still be ascertained as true or false. At the moment a judgement or proposition is ascertained as true or false it becomes a truth or it becomes a falsity. It is no longer simply a true proposition; it is a true proposition that is known as true. The true proposition has become a truth of the subject knower. In addition, the truth-value of mathematical truths is derived from the fact that mathematical truths are discovered as true. Thus, it is not the fact that a mathematical proposition is true that is the source of mathematical truth, for that would be to mistake the effect for the cause. The proposition or judgement is the means by which a truth is manifested; it is not the means by which the truth is manufactured. The expression 'mathematical truth' is favored because it demonstrates more clearly the relationship between the capacity of the human mind to discover or enunciate truths that are features of the universe and thus shows a closer relationship between epistemology and metaphysics.

3. Though Russell was a logicist, he also held that the individual simple propositions of arithmetic were self-evidently true on their own: 'All arithmetic, moreover, can be deduced from the general principles of logic, yet the simple propositions of arithmetic, such as, "two and two are four", are just as self-evident as the principles of logic.' *Cf.*, Bertrand Russell, *The Problems of Philosophy*, Oxford: Oxford University Press, 1943, Seventeenth Impression, p. 65.
4. This is what Fink means, it seems, when he says, with regard to phenomenological truths, 'There is thus no phenomenological understanding that comes simply by reading reports of phenomenological research; these can only be "read" at all by re-performing the investigations themselves'. Eugen Fink, *Sixth Cartesian Meditation*, Bloomington: Indiana University Press, 1995, p. 92.

5. It is this self-certifying character of philosophical truths that distinguishes said truths from the empirical truths of science which in contrast do depend upon confirmation by public and repeatable observation. It is the conflation of philosophical truth with empirical truth that gives rise to confusions in those like Latour who propose that truth depends greatly upon its acceptance by an established scientific community of observers. In Latour's words, 'In order for [truth] ... to be bellowed, hundreds of allies have to be present at the same time, allies about whom and about which rationalists are strangely silent ... Rationalists just confuse cause and consequence, the cheers of the victor with the army (e.g., the consensus of the agreed upon community of expert witnesses) that made the victor [truth] possible. They confuse what is held together - that is truth - with what is holding the whole lot together'. *Cf.*, Bruno Latour, 'Clothing the Naked Truth', in Hilary Lawson and Lisa Appignanesi, (eds.), *Dismantling Truth, Reality in the Post-Modern World*, London: Weidenfeld and Nicolson, 1989, p. 102. But, in philosophy an appeal to an outside community of observers to establish the veracity or authenticity of truth claims is not only irrelevant, it is counterproductive because it could assist in creating the impression that outside verification was a condition of truth value. In terms of this work, even the word 'outside' is a significant distraction. The truth experience is always inside consciousness. Its further discovery within an additional consciousness is not corroboration; it is another re-enactment of the truth experience. It neither adds nor detracts from the first. Just as 2 + 2 = 4 does not gain in truth upon its rediscovery even by the same consciousness, philosophical truth does not grow in value in direct proportion to the widening extent of its circle of knowers. Contrary to Latour, truth is the condition of acceptance, not the other way around. As Abraham Lincoln once said, one man in the right is a majority.
6. In this truth, one can find agreement not only in the conclusion reached, but also in the reasons adduced for the necessity of the conclusion between philosophies so seemingly far afield from each other as analytic philosophy and phenomenology. According to Bertrand Russell, with regard to the universal proposition, 'two and two are four', 'As soon as we see what the proposition means, even if we do not yet know if it is true or false, it is evident that we must have acquaintance with whatever is really dealt with by the proposition. By applying this test, it appears that many propositions that might seem to be concerned with particulars are really concerned only with universals ... It is quite unnecessary to know all the couples in the world: if it were necessary, obviously we could never understand the proposition, since the couples are infinitely numerous and therefore cannot all be known to us ... The general statement ... thus fails to make any statement about any actual particular couple. The statement made is about "couple", the universal, and not about this or that couple.' *Cf.*, Betrand Russell, *The Problems of Philosophy*, London, Oxford and New York: Oxford University Press, 1912, Seventh impression, 1976, p. 60. According to Eugen Fink, while he speaks in this context of

phenomenological predication, what he says would seem to be applicable even to the simple arithmetical proposition above: 'On the one hand, the natural meaning of the word and sentence points analogously to a corresponding transcendental sense, while, on the other hand, the intended transcendental meaning protests, as it were, against its expressional formulation, the sense to be expressed does not rest quietly in the expressional form, it is in constant rebellion against the constraint imposed upon it by the formulation in natural words and sentences. Thus all transcendental explications have a natural inadequacy, all concepts and sentences in one way or another fall short and in a particular sense fail before the demand that is, it seems, to be placed upon every predication (but especially the scientific)'. Eugen Fink, *Sixth Cartesian Meditation*, p. 88-89. Philosophy makes strange bedfellows.

7. It should be noted that mathematical truths are not only mathematical truths. For example, Euclid's tenth axiom in Book I of his *Elements* that two straight lines cannot enclose a space is utilized as an axiom in plane geometry. In this present, respective work, this truth is amended to read 'can *never* enclose a space'. (Of course, the space to which one is referred is two dimensional, but this is already understood from the use of the term 'straight'. In a curved space, two lines may enclose a space but the lines are then also curved.) In philosophical analysis, this selfsame truth, in its proper, philosophical formulation, is understood as an example of philosophical truth as well. From the standpoint of this present, respective work, there are no axioms. In this sense, the point of view expressed in this work differs slightly from Plato's description in *Republic* VI and the truths of mathematics do not rest upon assumptions any more than the truths of philosophy do. What was for Euclid an *axiom* is for the philosopher an epistemic truth, which is at the same time a metaphysical truth. The exact same data forms part of the subject matter of two disparate disciplines. Of course, the data serves entirely different functions *vis-à-vis* each respective discipline. (The preceding argument which refers to 'two straight lines cannot enclose a space' as Euclid's tenth axiom of Part I of his *Elements*, is not affected by the fact that according to T. H. Heath, Simson's edition of the *Elements* as revised by Todhunter deviates from the genuine text of Euclid as restored in Heiberg's edition. According to Heath, Euclid did not give as a common notion 'two lines cannot enclose a space' and this is only interpolated. *Cf.*, *Euclid's Elements*, Isaac Todhunter (ed.), London and New York: Everyman's Library, 1967, p. ix.)

Vivace, quasi allegro
Vivaciously and forcefully

2 On the Synthetic *A Priori*: Everything about the Synthetic *A Priori* Has Not Been Said

While the synthetic *a priori* may appear to be a most unfashionable subject, it is not to be believed that all that needs to be said about it has been said.[1] The prevailing custom is by and large to consider that the sorts of propositions that Kant considered to be synthetic *a priori* were in fact analytic propositions pure and simple, particularly the examples which he selected from the area of arithmetic. The notorious example of 7 + 5 = 12 is a case in point. For the moment, it is best to remove one's attention from this example, which, for all the thought that has gone into it, may no longer be capable of clear analysis. However, the fate of the synthetic *a priori* does not depend upon the validity of this particular example, which may perhaps be misleading and somewhat ill chosen as Kant's feature example.[2]

If, instead, one is to look at some other examples, the perspective of the synthetic *a priori* may be more clearly perceived. Take for example the proposition, 'whatever has shape has size'. It can be argued that shape contains within its concept the concept of size so that, this, too, is simply analytic. One need not psychologically think of size whenever one thinks of shape, or, if one were to, this would have nothing to do with the question at hand. It would only be a matter for empirical psychology. What is relevant is that the unpacking of the concept of shape logically discloses that size is contained, since shape has no logical meaning unless it includes size. Is this not sufficient to show the analytic property of the concepts? They need not be synonymous with each other and thus substitutable in expressions in order to display their analytic connection. One need not confuse the empirical habits of association with the logical implication of concepts.

As powerful at first as this analysis seems to be, it may rest upon a fundamental error. It is quite true that one need not appeal to the psychological habits of the thinker in question in order to attempt to establish the existence of the component of syntheticity. It also seems to be true that from a logical standpoint, the concept of shape is logically incomplete or incoherent without the inclusion of the concept of size. The real question, however, for the putative claim to syntheticity is *why* one must think of size whenever one thinks of shape.

Shape and size are not related solely in terms of a logical implication. That they are logically connected cannot be denied. But, the *ground* for the logical connection is very different from the ground for the logical connection that exists, say, between the concepts of 'biological mother' and 'woman who bears a child'. In the case of the latter two concepts, the two concepts are related because they are simply two different phrases which possess the same meaning (as stipulated by an arbitrary convention which is subject to cultural emendation or perhaps biological emendation). The logical relationship that occurs in the statement, 'whoever is a biological mother is a woman who bears a child' is purely linguistically grounded. The logical connection that occurs in the statement, 'whatever has shape has size' has nothing to do with the concepts of shape and size being synonymous and thus substitutable expressions. In fact, shape and size are hardly synonymous and substitutable expressions.

Whenever the word 'shape' is spoken, the speaker is obliged to think of size. The speaker need not be so obliged empirically, but the speaker is so obliged logically. *But, the logical necessity that obliges the speaker is not based on the meaning of the words.* It may be argued that size is a logical implication of shape, but the fact that it is a logical implication is not forced upon the speaker by the fact that the analysis of the concept of shape reveals that it implies the concept of size. This is mistaking the effect for the cause. The logical implication that exists between the two concepts cannot be denied, but the question remains as to why the logical connection exists in the first place. Why is it the case that, whenever one thinks of shape, by logical implication, one is compelled to think of size, or vice-versa?

The reason that one is obliged to think of size whenever one thinks of shape has nothing whatsoever to do with the conceptual meaning of shape. The concept of shape is not a pure concept in the sense that 'biological mother' is a pure concept. 'Biological mother' is a concept that is formed on the basis of linguistic convention. The concept of shape cannot be formed by the use of pure ciphers or signs alone. *The concept of shape can only be formed with the aid of something non-conceptual.* That non-conceptual element is something, which is and must be present in addition to that element which is purely thinkable. The non-conceptual element requires an act of visualization in order to bring it into being. The visualization of shape contributes an element which is not part of the conceptual meaning of shape. For lack of a better word, the convention of calling this non-conceptual element, 'space', is followed here. 'Space' must be added to shape in order to bring the concept into full focus. Whenever space is added to shape, the result is size. Whenever space is added to size, the result is shape. The 'sizeness' of shape or the 'shapeness' of size is in each case something which cannot be constructed on the basis of meaning alone. This component of the concept of shape or size is given to it by the addition of space. What the spatial component is, is not conceptually explicative. (It could not be, or else it would be a conceptual addition). But, it is this added element that is the ground for the logical necessity that exists between the two concepts of shape and size. Since this ground is not a logical ground, this can be called a synthetic element. In this fashion, a synthetic element is involved in the necessary connection that obtains between the two concepts.

The words 'synthesis' or 'synthetic' are perhaps not the most logically apt. The aspect to which this discussion points is not the act involved, or the psychological process required, in linking two concepts together. Such a synthesis is required even in the linking that must be performed between two analytically related concepts. The synthesis of which is spoken has nothing whatsoever to do with a psychological act. The certainty that accompanies the understanding that 'whatever has shape has size' has nothing whatsoever to do with the psychological state of the mind of the subject knower. While the subject knower possesses certainty in a psychological sense, that which is of interest is not this state, but how this state is obtained. A subject could be psychologically convinced (theoretically) that the moon is made of Danish Havarti cheese, but this does not affect the truth of the claim. The certainty which accompanies the act of knowing that whatever has shape has size is provided by the ingredient of the knowledge that is not conceptually derived. While this fact (that the ingredient is not conceptually derived or derivable) must be experienced, it does not follow that the experience of certainty to which one has access is only an experience of certainty in the psychological sense of the feeling of certainty. This feeling of certainty is a necessary accompaniment to the moment of the understanding of the truth of the relationship that exists between the two concepts of size and shape, and it occasions wonder about the basis for this certainty.

The relation of father to son is a relation that does not require the intervention of space for its intelligibility or truth-value. It is a relation that is singularly based on the pure meaning of the concepts. Synthesis, in the sense which is given to this term in the present context, has nothing to do with putting two concepts together. Syntheticity has to do with the fact that something is given which is not present in the bare concept. This is what Kant, were he to have been a self-conscious pure phenomenologist, would or should have said. His examples and his method of explaining his examples have to a large extent distorted the nature and the brilliance of his discovery that there are, in fact, two sources of knowledge, only one of which in a preliminary sense, may be said to be conceptual. (It is not clear that Kant himself always understood his discovery in precisely this fashion.) Kant's task was complicated by the fact that he was also at the same time endeavoring to explain how one could account for the sense experience of the empirical world. If one exempts the pure phenomenologist from this latter task, then the pure phenomenologist is under no obligation to show how processes, whether transcendental or psychological, have something to do with constituting the empirical world.

All that is required of the pure phenomenologist *qua* phenomenologist is to elucidate the distinctions that appear among the concepts and the objects that appear in pure consciousness. In this sense, it would appear that there are necessary relations between the data of pure consciousness that are not based upon the analysis of the meaning of words alone. The meaning of 'shape' requires a non-conceptual element for its cognition and that non-conceptual element is not and cannot be conventionally stipulated. The non-conceptual element is not invented or contributed by the subject knower. Kant was quite right to criticize the power of pure reason. The appropriate direction of the critique must now be understood to be in that direction in which pure reason cannot operate with pure concepts alone. It must take into account the contribution of another source of knowledge in the description of the phenomenology of mind. It is not that the subject knower must forever be restricted to knowledge of phenomena (which is the direction taken by Kant). But, in the knowledge of the data of pure consciousness the subject knower must attend to different types of available data. It is in this sense that the synthetic *a priori* remains a topic about which everything that can be said has not yet been said.

To sum up the meaning of this chapter: what is at issue is to clearly understand the distinction between the necessary connection that obtains when two words have an identical meaning, and the necessary connection that obtains when two concepts are related because of their common sharing of a third element which is non-conceptual.[3] In this respect, one may also be psychologically certain that bachelor and unmarried male possess an identical conventional meaning (when it is stipulated that the term 'bachelor' is to be reserved for those of the male sex), but the identity of the conventional meaning of the two terms is not based upon the psychological feeling of certainty that attends the moment of knowing or realizing their identity. If there is a psychological feeling of certainty that attends the latter moment of knowing, it is based upon an empirical assurance of knowing that these two words stand for the identical concept. The feeling of certainty that attends the former moment of knowing (e.g. in 'whatever has size has shape') arises from a certainty that attaches to the moment of cognition itself. It is necessary to distinguish between the kind of certainty that characterizes 'can never be otherwiseness' from psychological certainty, since the latter might be a form of certainty to which a delusional psychotic may lay claim.

Notes

1. For example, while one can find considerable discussion of the concept of the synthetic *a priori* in Paul K. Moser, (ed.), *A Priori Knowledge*, Oxford: Oxford University Press, 1987, there is no discussion of the peculiar basis of *a priori* syntheticity which is advanced above. In Roderick M. Chisholm's chapter, 'The Truths of Reason', he offers examples of synthetic *a priori* propositions such as 'Everything red is colored', and suggests while these express what is known *a priori*, they cannot be shown to be analytic. *Cf.*, p. 141. However, in terms of the view being expressed in this present work, such a proposition would not be construed as synthetic since red is a species of the genus color and could be thus displayed as analytic. In terms of the most common arguments advanced contra the synthetic *a priori*, it would seem that Lewis White Beck's comment found in his essay, 'Can Kant's Synthetic Judgments be Made Analytic', is still relevant: 'The real dispute between Kant and his critics is not whether the theorems are analytic in the sense of being strictly deducible, and not whether they should be called analytic now when it is admitted that they are deducible from definitions, but whether there are any primitive propositions which are synthetic and intuitive. Kant is arguing that the axioms cannot be analytic, both because they must establish a connection between concepts, just as definitions do, and because they must establish a connection that can be exhibited in intuition. And this is what the modern critic of Kant denies. I think Kant is obviously right in saying

that there cannot be a system of nothing but analytic propositions; there must be some complexes to analyze, and these must be stated synthetically'. *Cf.*, Moltke S. Gram, (ed.) *Kant: Disputed Questions*, Chicago: Quadrangle Books, 1967, p. 245. Kant himself may have contributed to some of the confusion surrounding this issue with respect to his justifications for the distinction between synthetic and analytic judgements which are not to be identified with the justifications advanced in this present work. In his *Lectures on Logic* and in *On a Discovery*, Kant distinguishes analytic judgements as extending knowledge in a formal sense and synthetic judgements as extending knowledge in a material sense. However, from a phenomenological point of view, such a distinction need not be entertained. The distinction as argued in this present work relates to a distinction in different sources of knowledge, one of which is configurative or spatial and the other of which is intellectual or semantic. *Cf.*, Henry E. Allison, *The Kant-Eberhard Controversy*, Baltimore and London: The Johns Hopkins University Press, 1973, pp. 44-5 *et passim*.

2. In the fourth chapter of his *Language, Truth and Logic*, A. J. Ayer pointed out that the justification which Kant gave for '7 + 5 = 12' being synthetic was psychological whilst the justification he gave for 'all bodies are extended' was based on the principle of contradiction whereas, according to Ayer, both judgements rest on the principle of contradiction and therefore were both analytic. But if one considers Kant's example of a synthetic proposition to be that a straight line between two points is the shortest line, one is on more secure grounds. Here, as Kant argues, 'straight' (a quality) is by no means contained in the concept of 'shortest' (a quantity). *Cf.*, Introduction to the *Critique of Pure Reason*, V. Kant then states that, 'Intuition, therefore, must be called in; only by its aid is the synthesis possible' (all English translations of Kant are by Norman Kemp Smith unless otherwise stated). From the viewpoint of the present, respective work, it is not the work of intuition as a *noetic* function that is relevant (which Kant probably did not mean), but that one must go through the medium of space in order to see the truth of the judgement. In other words, it is not that one must have recourse to 'intuition' as non-logical seeing, but it is the case that the intervention of space is necessary in order that the judgement can be seen to hold. One is aware of the truth of the judgement in a non-conceptual fashion, which involves a type of brute 'seeing' or connectiveness. The brute 'seeing' is the *effect* of the fact that the truth arises from the non-conceptual character of space; it is not the cause of the kind of knowledge that one possesses or the reason why one possesses the kind of knowledge that one possesses. In order to become aware that two straight lines can never enclose a space, one must visualize this truth fact in order to become aware of its truth. Without visualization there is no way to know the truth to be true. It could be objected that this is a truth that follows from the definition of straight lines, but there is no way to know that this is true apart from the test of visualization. The question is, how is it known that this is a definitional truth (if it is)? It is only known

to be true - whether definitionally or not - through the medium of space. The truth arises from the nature of space. Otherwise it could never be known to be true nor would it exist as a truth, definitional or otherwise, in the first place. That it must be visualized in order to be seen as true is due to the fact that its epistemic character possesses no conceptual origin, but is due entirely to the nature and the existence of space. Kant advances a similar argument in his *Inaugural Dissertation*. Section III. #15, C. *Cf.*, *Kant's Inaugural Dissertation and Early Writings on Space*, John Handyside (trans.), Westport: Hyperion Press, 1988, p. 60. Kant puts the point in a slightly different fashion in B268 of his first *Critique* where he states that, 'Thus there is no logical contradiction in the concept of a figure which is enclosed within two straight lines since the concepts of two straight lines and of their coming together contain no negation of a figure. The impossibility arises not from the concept in itself, but in connection with its construction in space, that is, from the conditions of space and of its determination'. In the point of view of the present, respective author what is emphasized is the non-conceptual character of space as a mode or source of knowledge.

3. While irrelevant to the argument advanced in this chapter, it is of interest to note that even purely analytically true sentences are not merely certainties but necessities and this carries implications for natural languages as not being merely conventional. It is not merely that one is certain of their truth (which is a psychological fact), *but that they cannot not be true (an epistemic fact)*. Of course it can be argued that it is convention that sets up tautological meaning. But, once a language system is constructed, it is impossible not to be able to construct tautologies. If a language system cannot help but contain the possibilities of tautological truth, then is it possible that linguistic systems are purely conventional? This appears to be the argument that is made by Paul M. Postal and Jerry Katz. *Cf.*, Geoffrey J. Huck and John A. Goldsmith, *Ideology and Linguistic Theory, Noam Chomsky and the Deep Structure Debates*, London: Routledge, 1995, p. 136.

Allegro moderato ma deciso
Quite lively and with decision

3 Kant's Four Missing Arguments: A Phenomenological Treatment of the Kantian Idea of Space

Kant was interested in providing the foundations for empirical knowledge and mathematical knowledge. Kant started with human knowledge - or what he assumed was known - and then asked what were the transcendental conditions for the possibility of that knowledge. This is at least what seems to be his philosophical method from the order of explanation followed in his first *Critique*. It can be said that he possessed philosophical courage to begin with what he took to be knowledge. It can also be said that this was philosophical cowardice, because he had begun with what was taken to already count for knowledge by other disciplines instead of starting out in totally unknown territory. In any case, he came to no harm with his beginning with mathematical knowledge; it was only in his acceptance of co-temporary truths of physics that his knowledge system has had occasion to be misunderstood.

Hegel's definition of space as pure externality, or the outsideness of things to each other, appears to bear at least some resemblance to Kant's first argument in his metaphysical exposition of the concept of space.[1] Outerness, for Hegel, is not the same as outsideness in the sense of outsideness as over against insideness. Outerness is the externality of outer things to each other. Insideness could then be taken to be a sub-species of or a kind of externality although it is not to be suggested that this was what Hegel intended.

Hegel's definition of space is more closely correlated with Kant's discussion as it appears in the Aesthetic than with Kant's discussion as it appears in the Dialectic. This in and of itself is an interesting development and one which merits further investigation.

In a sense, the work of the present author may be traced back to a sequence which follows directly after Kant, but which is armed with the knowledge of phenomenology as the domain of interest as pointed to by Husserl. For the development of the topics treated in this present work, it is more to the point to trace the origin of the concerns of this present work directly back to Kant. But unlike Kant, the primary interest here is not in elucidating the conditions for empirical knowledge. While Kant initially appears interested in showing the conditions for mathematical knowledge, this interest is too soon left behind. This is rather a pity, for a deeper investigation of the relationship between mathematical knowledge and philosophical knowledge would shed a greater light on philosophical knowledge, since the ways of truth in these two disciplines are more closely akin to each other than the ways of truth in philosophical knowledge and the empirical sciences. In addition, there does not appear to be an investigation of extraordinary knowledge, the kind of knowledge that is not empirical and cannot be explained empirically, such as the knowledge possessed by *idiots savants*. What are the conditions for the possibility of the existence of such knowledge? While Kant does not take up such questions, these questions are discussed in the present author's, *A Metaphysics for the Future*.

If, for a moment, one considers Kant's fleeting but brilliant treatment of mathematics, it is obvious that by Kant's own definition, mathematics is not a species of empirical knowledge. Thus, by pointing to the transcendental conditions for truth in mathematics, Kant reveals that there are conditions that enable human beings to know that which cannot be explained by empirical means. Mathematics consists of the kind of knowledge that cannot be obtained through the senses. This in and of itself is transcendental knowledge in the sense of the term 'transcendental' utilized in this present work, and, so it would appear, for Kant as well in respect to his usage in his first *Critique* (B25). In showing the transcendental conditions for the possibility of mathematical knowledge, Kant points the reader in the direction of mathematics as a kind of transcendental knowledge (knowledge which cannot be obtained by empirical means).

In the twentieth century there was an attempt to move mathematics out of the transcendental sphere, and reduce the truths in mathematics to matters of mere logical definitions, so that there was insufficient effort made to develop Kant's suggestion that there are two sources of knowledge. Before re-entering the metaphysical arena of space and time, it may be of some interest to examine Kant's metaphysical exposition of the concept of space in his *Critique*, not in terms of what Kant thought it was establishing, but in terms of what progress he may have been making, perhaps, unbeknownst to himself, in terms of fundamental epistemological distinctions in phenomenology. This examination will focus specifically on the contributions that his arguments have made for the development of phenomenological epistemology. The examination which follows does not in all respects follow Kant's meaning but represents a phenomenological purification of Kant.

Kant's first argument is that he cannot represent anything as outside of him or represent appearances as alongside or outside of one another without first utilizing the concept of space. This argument has been much maligned as many have responded with the Lockean-like reaction that outsideness may be learned empirically. Why does outsideness need to be presupposed beforehand? The argument, brief as it may be in Kant, is perhaps too profound to be understood in its simplicity. It may be of assistance to present Kant's argument from a phenomenological perspective. It is important to bear in mind that this perspective is not in all respects identical to the perspective which Kant himself was aware of taking but it is a perspective which is compatible with much of what Kant says and, indeed, from which most of what Kant says can obtain a more consistent and fruitful meaning.

In phenomenological terms, outsideness cannot be learned because outsideness is a subjective perspective. Outsideness is always relative to oneself. One cannot learn what is outside *of oneself* from that which is outside. Outsideness is only experienceable from the inside out. If an infant reaches out and touches something and realizes it is something other than herself or himself, the infant must first have a concept of herself or himself in order to distinguish herself or himself from the other as other. Obviously, this concept of self can only be experienced from the inside. But, if one only possessed access to that which was on the inside, how could one ever get to the outside? *Outsideness only makes sense phenomenologically as an inner distinction that is made prior to so-called outer experience.* Space, which is what makes possible the distinction of inside-outside, must already be known before one could know that something could be outside.

To put this matter in another way, how could one be aware of an inside perspective unless one already possessed an outside perspective?

The notions of inside and outside are correlative. One could not grasp what was outside unless one were already acquainted with what was inside. But one could not know what was inside unless one already knew what outsideness was. On the other hand, one could not know the meaning of outsideness unless one were already acquainted with what was inside. Thus, the two notions must be possessed together. The inside-outside distinction, or to put this in another way, the capacity for spatial representation (one must desist from the temptation to think of space as an empty box into which one fits things) is not learned from empirical experience.

That Kant's argument does not appear to read this way is perhaps due to the fact that in his first argument, Kant appears to make use of an 'omniscient eye' perspective rather than a subject's point of view perspective. In other words, he argues as if he were viewing himself – by which he seems to mean his body – from a witness standpoint that possesses no specific location. That Kant is making use of an 'omniscient eye' is suggested by his avoidance of the terms 'inner' or 'inside' and his parenthetical explanation of 'outside me' as '... something ... in another region of space from that in which I find myself ... ' thus conflating outside [of his body] with alongside. That this interpolation of Kant is plausible is strengthened by his continuing after the parenthetical explanation to say, ' ... in order that I may be able to represent them as outside and alongside one another and accordingly as not only different but as in different places ...' thus creating the impression that by 'outsideness' he mainly intends 'outerness'.

For Kant, the argument is made by referring to outsideness or alongsideness. In the first argument, Kant does not explicitly refer to insideness because for Kant, time is the form of inner sense, i.e., the intuition of ourselves and of our inner state (B49). However, tacitly Kant cannot avoid making reference to that which is inner because even for Kant the reference to that which is outer is explained by contrasting the outer with a self-reference (B38). In this reformulation of Kant, the subject knower must already possess the concept of inner-outer before the subject knower can know that any experience is outer. It will be easily granted that the subject knower must know what is inside from the inside. What is not so easily understood is the fact that if the subject knower were already to know what is inside, then even to know this, the subject knower must already have a distinction between inner and outer. *Inner and outer are distinctions within consciousness, not distinctions between an interior world and an exterior world.*

Hegel's famous definition of space shows what he learns from Kant, though he applies it in a non-phenomenological direction. For Hegel, space is the outerness of objects to one another. If this is understood and applied phenomenologically, then the outerness or outsideness of which Hegel speaks is a distinction that is made inside of consciousness (this is not to be confused with Hegel's intended meaning). One could, in Kantian terms, judge that something is outside of oneself (using one's body as a frame of reference), only if one first understands what is meant by outsideness. In phenomenological terms, Hegelian pure externality or 'outsideness of one another' is a distinction within consciousness and not between external objects. The externality that exists between external objects is only one case of externality. (There can be externality between inner objects as well.)

What must be borne in mind is that the inside-outside distinction is a distinction among the data of pure consciousness and has nothing to do with the space of the physical world. The mental capacity to place two blocks next to each other in one's imagination cannot be learned from an outside example. To place the two blocks next to each other in one's imagination requires a spontaneous mental activity. This activity cannot be learned from the outside any more than the ability to read can be learned from the outside. The seeing of two blocks alongside one another in the empirical world is no different from the act of seeing them alongside one another in one's imagination. The 'outerness of the blocks to each other' cannot first be learned from empirical experience, and then applied to imaginative experience. It is the very same imaginative power at work in both cases.

Whether it were imagination or empirical experience that one first happened to possess, the imposition of an internal concept of experience is the condition for the possibility of the experience. However, it would seem that the internal capacity for placing two objects alongside one another in one's imagination is logically prior to the capacity of seeing two objects next to each other in the empirical world, for the imagination cannot learn its capacities from outside examples unless it already possessed the power of such knowledge on its own.

Kant's second argument has been much criticized as being of only psychological relevance. It has been objected that the putative failure to empirically think space away has to do with the limitations of human psychology or with the impoverishment of one's imagination, and nothing follows from this as to the *a priori* character of space. In a way this is so, but it is important not to neglect this argument for it contains the same truth that is utilized in the first argument. The fact that one cannot *think* space away is an illustration of the non-conceptual character of space (in the phenomenological sense of the meaning of space). Even the notorious allegedly impossible act specified by the dictum, 'don't think of a pink elephant', is an empirical possibility. It is possible that the mind can be made a blank for at least a finite period of time. But this 'blank', while empty of objects, (despite Husserl's ideas to the contrary), is nonetheless a 'blank' field of extension - or space. That this field of extension cannot be thought away proves that it is not a creation or a resultant of thought. As a result, Kant's second argument, however much maligned, is in fact a brilliant proof that there is a mental datum which is not conceptual and thus is already suggestive of the existence and the character of a second source of knowledge.

At the same time, if Kant's 'argument' is a valid phenomenological test, it demonstrates that there can be an objectless consciousness and thus presents itself as a refutation of Husserl's claim that consciousness is always of an object. Kant's second argument is a phenomenological experiment. Kant thought that he could experience consciousness as objectless. If one could not think the objects of consciousness away, perhaps that represents a failure of the imagination.

All concepts may be thought away, but what is left over is something, though it is not a discrete object. What is left over is a primitive residuum, which is not conceptual, because it cannot be thought away, wheras all thougths (despite the 'pink elephant argument') can be thought away. (This is an empirical question as it is possible in principle to think all thoughts away.) The primitive remainder is a reflection of the fact that the pure data of consciousness do not include only conceptual objects. What is left over cannot be thought; but what is left over is not nothing at all. By whatever name it is called, its nature is that of some kind of primitive existence which is stripped of all conceptual characteristics. This residuum may be that to which philosophers made reference when they devised such seemingly abstruse philosophical terms as 'thusness' or, 'isness'. It is just there, in its dumb, characterless state, whatever it is. It is thought's leftover. This primitive residue or 'leftoverness' is an existence which is a 'just thereness'. It is not the case that nothing is there. The mistaking of 'just thereness' with nothingness has perhaps been the source of many a philosophical dilemma concerning the so-called problem of nothingness. What is the case is that what remains when thought disappears is an existence which always remains as a pure residuum of consciousness. Since it cannot be given any conceptual form, it must be designated as pure, primitive, conceptually barren existence. This pure existence is a primitive form of distinctness that differs from the distinctions that one later (or independently) discovers by concepts to obtain between concepts. This primitive form of distinctness can now be designated by and dignified with the proper term, 'space'. When philosophers in the past referred to 'isness' or 'thusness' or 'existence' or 'being in general', they were attempting to articulate the concept of space, but could only do so with a lisp. It is not surprising that Hegel argued that Being in General was Nothing in Particular, a description which unintentionally captures much of the essence of the nature of space.

In Kant's third argument, when Kant says that all spaces are parts of one space, he can be interpolated as saying that there is no conceptual distinction between the contents of one non-conceptual distinction and another. *The pure form of difference is conceptually undifferentiated.* This form of difference can and must only be experienced phenomenologically. But the distinction which is experienced cannot be conceptually drawn. As argued in the present author's *A Metaphysics for the Future*, this is the truth behind Leibniz's famous principle of the identity of the indiscernibles. The distinction between one and one and one and one can only be made intelligible in terms of a difference in instantiation and this may only be represented spatially or temporally and not conceptually. Kant's argument is important in that it makes clear that there is another form of distinctness other than that form of distinctness which is formed by analytic distinctions between concepts. The character of this non-conceptual form of distinctness will be further enlarged upon in the sequel.

Kant's fourth argument is that while empirical concepts can possess instantiations under themselves, spatial distinctions can only possess existences within themselves. Thus, while all tigers do not exist within the concept of tiger, all spaces are parts of the selfsame space. This is an important source for understanding the distinction between conceptual distinctions and a primitive, non-conceptual form of distinctness.

Despite post-Modernism, an empirical concept cannot be indefinitely complex or it would be devoid of meaning. Space, however, is capable of being infinitely divided within itself. This is how Kant's account of space as an infinite given magnitude can be interpolated. What is important for the present purposes is not the infinite divisibility of space. What is important is that the primitive form of difference is not analyzable but at the same time is experienceable. As in the third argument, one moment of distinctness is only different numerically or spatially from another. This is only possible to be apprehended in a state of pure consciousness devoid of concepts. To properly understand Kant's four arguments, one must adopt the perspective suggested in his second argument. Kant's second argument can be characterized as a recommendation for the reader to perform a thought experiment which has as its objective the achievement of a limited version of Husserl's *epoché* with a special emphasis on the elimination of conceptual thought. Space and time are primitive (by that is meant irreducible rather than less sophisticated) forms of distinctness or, in Plato's precise and illuminating discrimination in his brilliant dialogue, *Parmenides*, 'pure forms of difference' as opposed to 'otherness'. What Kant thinks of as the infinite given magnitude of space, thus infinitely divisible or extensible, is the character of space as being incapable of being distinguished within itself since it is an irreducible (even for itself) form of distinctness.

Thus, Kant's arguments have now been phenomenologized. It is not being argued here that space is the condition for the possibility of empirical experience. It is perhaps even a mistake to continue to use the word 'space' to refer to this elemental mode of distinctness. It may be that Plato's choice of the term 'difference' in *Parmenides* might be more apt although it may be too wide as is argued in the sequel. In any event, it is difficult to avoid or resist the rich philosophical inheritance that the usage of the term 'space' provides. It is only important to keep in mind that this is not physical space in any sense. If one were to revise the objective of Kant's metaphysical expositions of the concept of space, one could say that they demonstrated *a priori* that there must be a form of distinctness, which is other than conceptual. *These four arguments are, though Kant does not explicitly recognize them as such, properly speaking, Kant's missing arguments which demonstrate the existence of a second source of knowledge.* It is best to understand these four arguments as pure descriptions of the phenomenological form of distinctness that may be referred to as 'space'.

The foregoing may be taken as a phenomenological purification of Kant. Kant was perhaps too much taken with the great achievements of empirical science of the Modern age to be a pure phenomenologist. He was, after all, chiefly interested in showing the conditions for the possibility of *empirical* experience. In addition, Kant wanted very much to show that space and time were somehow forms of human intuition (in his sense of 'intuition'), at the expense of physical space and time altogether. While these interests were counter-purposive, they also have made the real achievements of Kant less discernible and less compelling.

In pure phenomenology, there is no interest in inferring that space and/or time are forms of intuition. In making some such claim one is always vulnerable to the charge of so-called faculty psychology.

Today, such questions are far better addressed by the cognitive psychologist and are of no immediate interest to the philosopher. (The relation between philosophy and the empirical sciences is discussed below in Chapter Seven, 'The Analogies of Experience'.)

Similarly, in pure phenomenology, no claim is being made about the nature of space and time as they exist in the world of the physicist. The pure phenomenologist is only interested in what may be referred to as phenomenological space and phenomenological time. These are, for phenomenologists, forms of distinctness that are made, nay, in fact, must be made, within consciousness. Perhaps, it is a mistake to use terms like 'space' and 'time' that carry the same name and connote such different possible meanings. But, in Chapter Seven of this present work, perhaps this use can be justified by indicating how the physical world can serve as a mirror to enable the subject reader to understand the world of the phenomenologist and thus justify the use of these terms in common.

This present chapter can be concluded with a final word concerning faculty psychology. From the standpoint of pure phenomenology, one makes no claims as to the existence of any special faculties or a mind in which these faculties inhere. One is concerned only with the pure data of consciousness and describing the relations that obtain between and among the pure data. In this respect, to some extent, the pure phenomenologist is closer to the transcendental philosophy of Hegel than to the more empiricist-flavored philosophy of Kant. On the other hand, Hegel diverges from phenomenology, despite his choice of that word to form part of the title of his unique and unforgettable work, in that he is too fixed on showing some special development that obtains between these concepts (a dialectical one), which borrows from theory and diverges from the task set for pure phenomenology, which is pure description of what is found, not invented by or constructed in theory.

In carrying out the task of pure description there is no need to justify the truth of claims by appeal to correspondence to fact, or to the coherence of any world-system that is being constructed. One only needs to ask that each reader carry out the work of introspection from the standpoint of pure consciousness. The ultimate court of appeal for the truth of the statements to be found within this section is the pure experience of the subject reader. The pure phenomenologist begins from pure experience, and never leaves this frame of reference. In no sense, then, is this purely speculative or conjectural and thereby ungrounded metaphysics. Each reader is called upon to test every claim made within this section within the crucible of her or his own phenomenological experience. If whatever is claimed to be necessarily true cannot be experienced in pure consciousness, then it is reducible to armchair theorizing with which this work is not concerned. In this sense, the standpoint of this work is more radically empiricist than the empiricism of the empiricists, for sense-experience is always open to doubt and to questioning for its authenticity. In this respect, the work of philosophy is more like that of the bee in Bacon's famous analogy, than that of the spider or the ant. But within this analogy, the philosophical bee is a bit closer to the spider than to the ant. For the moment, to revise slightly Kant's famous statement: while all that is known begins in experience, all that is known also comes from deep conscious experience. The experience of the subject knower is much deeper and more profound than sense experience. Kant was led to make his famous statement, because he thought that one could not know that which was certain from experience. That is true, of course, if experience is understood in the Humean sense. One can never know with certainty anything that is experienced from one's senses. Great insights may come to one in a sensuous frame, e.g. in words, but the content of the knowledge is not reducible to its sensuous frame and the source of the knowledge also differs. The means by which one is aware of the knowledge of the sensuous frame may, for example, be through hearing, but hearing is not the source by which one may become cognizant of the contents of the knowledge.

In the area of pure phenomenology, the claim of the phenomenologist is both always made with certainty, and also from experience, but it is from pure experience that the phenomenologist proper speaks. Phenomenological experience may then be utilized as a base from which analogies to a higher order of reality may be inferred, but this level of metaphysical thinking can only be properly conducted when it is grounded upon phenomenologically discovered truths.

Note

1. While Kant is well known for holding that time is the form of inner sense, it would appear that the concept of outer does not make sense without a prior understanding of inner. In this sense, although Kant does not say this explicitly, inner and outer can only be correlatively understood. In this regard, it seems that space is more primordial than time (it is interesting that Kant first treats space and then time), for even if time is the form of inner sense, the very understanding of inner is not possible if one does not first understand inner-outer, which requires the form of space. In Kemp Smith's translation of Kant's relevant point in his first *Critique*, 'For in

order that certain sensations be referred to something outside me (that is, to something in another region of space from that in which I find myself) ... the representation of space must be presupposed' (B38).

Andante con moto
Rather slow and sustained

4 Space, Time and Metaphysics

Space and Time

It may appear that already too much attention has been paid to space and time, which are surely highly specialized topics, and that in a work that pretends to any profundity, one should proceed with the discussion of broader and weightier themes, and not focus so much on such narrow and relatively insignificant themes. It is, however, perhaps, precisely because space and time have been brushed aside with such little attention as either belonging to the technical provinces of the theoretical physicist, or, in any case, as too unexciting or too narrow for the philosopher to say that much about, that greater philosophical problems have developed.

The importance of treating space and time in such depth and to such an extent is that all appearances can and must make their entrance in the theater of space and time. For the phenomenologist this is of special significance. If the phenomenologist is only interested in what appears in consciousness, and all appearances which appear in consciousness can only appear when represented spatially and temporally, then the nature and the existence of space and time assume a special importance.

Space and time are concepts which must be treated together because they are interdefinable. This is the beginning of a definition of space and of time, which is quite unlike the definitions one finds in Aristotle, Kant or in any other previous philosopher. Space cannot be defined without reference to time, and time cannot be defined without reference to space. Both are, in a sense, primitive terms, in that no other concept can be employed that would be adequate to define either one or both. But, space or time are in fact utilized in the making up of all other conceptual definitions.

Strictly speaking, space and time are not concepts, as Kant well noted in his *Critique of Pure Reason*. In the sense in which the terms, 'space' and 'time', are being employed in the present work, they are not concepts because they are required in the composition of all other concepts. For Kant, 'space' and 'time' are what he refers to as forms of sensible intuition. In the vocabulary which is more apposite for the present work, space and time are Forms of Plurality. The use of 'Plurality' is to be preferred to Plato's discussion of 'Difference' in his *Parmenides* since there are different forms of difference and 'Plurality' is used herein to refer to non-conceptual distinctness.[1] Space and time are the twin conditions for the possibility of multiple representation. If, for convenience's sake, there is employment of the terminology of the concepts of space and time, then such a use of vocabulary is in the manner of Kant. The word 'concept' is only a conceptual fiction, a shorthand referral mechanism to the phenomena, which are space and time. 'Space' and 'time' are not themselves ideas. One must have access to a mode of plurality which is not obtained through concepts in order for the possession and the comparison of ideas to be possible in the first place. Space and time may be initially spoken of as the forms of individuation and pluralization.

No one, of course, really possesses an idea or a concept of space or time. To make reference to the concept of space or time is only a manner of speaking, a *façon de parler*. If someone were to ask what was space or what was time, it would be impossible to point to any illustration taken from the empirical world, in the same way that one could illustrate an empirical concept, such as 'dolphin', by pointing to examples taken from the empirical world. But, no empirical example of space or time can be pointed to in such a way as to clearly demarcate it. For Kant, any example of space or time was a portion of space or time, and that was one way to distinguish space or time from empirical concepts, since every representation of an empirical concept was not a portion of the concept. However, in the frame of reference, which is employed herein, there cannot be any such thing as pieces or portions of space or time. To refer to either pieces or portions of space, or to refer to the infinity of space, is to make reference to space or time as if one were referring to a physical object - and despite his desire to avoid such a consequence – proved to be an implication which Kant himself did not prove entirely capable of avoiding. Since the argument from parts to whole and the argument from the infinite extensibility of space and time are key arguments in Kant's distinction of space and time from concepts, Kant cannot then say that space and time are only subjective frameworks. A subjective framework is neither susceptible to part-whole nor finite-infinite analysis.

From the standpoint of this present work, space or time are the frames in which there are any pieces or portions or divisions or additions whatsoever. They cannot, therefore, be themselves pieces or portions, divisions or extensions in any sense. Thus, no empirical example of space or time can ever be pointed to as a part of space or time. While it is strictly speaking illegitimate to apply the attributes of infinite divisibility or infinite extensibility to space and time, one cannot by fiat forbid such application. It is important to remember that whenever one applies the attributes of infinite divisibility or extensibility to space or time one makes use of the Forms of Plurality.

It may be asked why can it not be said that space and time are metaphysical concepts rather than empirical concepts? Why should such Platonic and Kantian language as Forms (albeit not forms of sensible intuition) be employed, when a more traditional device such as a metaphysical concept might be utilized as a designation? The consideration of space and time as Forms of Plurality, rather than as metaphysical concepts of some kind of another, is because a metaphysical concept could be understood as some kind of basic assumption that one makes in order to explain the rest of the world (as in the case of a cosmology), or in order to deduce in some way the rest of a system of concepts (as in a rationalist metaphysics). Space and time are not assumptions that one makes in order to explain in some sense the rest of the world (as for example, the origin of the world), since there is insufficient evidence at the assumption stage to argue that there is an inherent connection between a metaphysical concept and the rest of the world that is explained by it. For example, there is no inherent connection between a Creator of the world and the world that She or He creates. Neither are space and time fundamental presuppositions from which one may deduce in some fashion a system of less fundamental ideas. There is no deductive relationship between space and time and other concepts.

There is no escaping the categorization of space and time as Forms, no matter the outcry such an old-fashioned designation might arouse. By 'Form' the meaning intended by Plato, is not intended. By 'Form' nothing other than frame is intended, except that the word 'frame' carries with it no traditional history. In addition, the word 'frame' is itself limited, in that in order to have any frame, one must already have a form of plurality. One could, perhaps, restrict the word 'frame' to stand for the ultimate mode of making up plurality, if all that is under consideration is the basis of a choice of vocabulary. But, more important than the actual terms used is the reason why it is said that those terms possess the meaning that is attributed to them.

For the time being, then, it is valuable to follow the custom of using the term 'Form', if for no other reason than to make it very clear that no ordinary metaphysical or empirical concept is being employed. But, a more important reason for the retention of philosophical vocabulary is that it may provide a reason to investigate more fully why certain terms were used in the past, and more importantly, to investigate more fully the aspects of reality pointed to by the use of such terms. The abandoning of such traditional vocabulary, contrariwise, may lead future philosophers to abandon or at least neglect the study of the past, a fate that can condemn philosophy, like its sister, history, to repeat the past. Such was the fate of much of twentieth-century Western philosophy, which repeated the philosophy of the fourteenth and the eighteenth centuries. (The question may be asked if philosophy is significantly deteriorating since it required only two centuries to return to the thought of a previous time whereas earlier it required four centuries of progress before a retrogression to the previous level of philosophizing occurred.)

Before proceeding, a very legitimate concern must be addressed. Everyone has some popular acquaintance with space and time. Quite obviously, they play a very large role in one's everyday life, not to speak of the very important role they play in the world of physics. Why is it that the philosopher insists on speaking about space and time in some other, extremely counter-intuitive or at least counter-empirical sense?

Whatever is said about space and time has no immediate application to the popular understanding of these concepts, nor does it have a very immediate application to the understanding of these concepts as physicists employ them. There may be some very interesting intermediate applications that may be relevant to the world of physics. At the outset, however, such concerns are to be set aside.

The point is that there is certainly a legitimate use of these concepts as they are of crucial importance in both the everyday world and in the world of the physicist. However, this should not blind the subject inquirer to the fact that there is another, and equally legitimate use of these concepts which has a relevance on the transcendental plane. For instance, it is of importance to note the relationship of the Forms of Plurality to concept formation.

For Kant, since there can be one and only one space (and one and only one time), and space and time are forms of intuition, then it would seem that there would no longer be the space and time of the physicist. In his first *Critique*, Kant wrote, '... if the subject, or even only the subjective constitution of the senses in general be removed, the whole constitution and all the relations of objects in space and time, nay space and time themselves, would vanish' (B59). And again, with respect to time, 'If we take away from our inner intuition the peculiar condition of our sensibility, the concept of time likewise vanishes; it does not inhere in the objects, but merely in the subject which intuits them' (B54).

While the problem of the relationship between the everyday use of the concepts of space and time, and the philosophical use is not easily solved, some preliminary suggestions can be made. The Forms of Plurality are the transcendental conditions in Kant's sense of the everyday use of space and time. Perhaps, it is a mistake to discuss even the Forms of Plurality under the labels of space and time. But, given the connection between the Forms of Plurality and space and time, it would be difficult, if not impossible, to continue to avoid this association indefinitely. The common-sense usage of the terms 'space' and 'time' is dependent upon the fact that one thing can be distinguished from another in the first place. The space and time of the physicist, or the person in the street - the space, of which, for example, the physicist in a philosophical mood might ask, is it infinite or finite? - is not the same application of the concept as that which the phenomenologist renders of it. (It must be said that even in contemporary physics that the concepts of space and time are far from clear. For example, it is currently held that space and time are finite but unbounded. This, however states a problem as a solution. If an entity is unbounded, by what right can one state that it is finite?)

If one were to make a distinction between the popular and the phenomenological use of the term 'space', it would not then follow - as perhaps Kant feared - that there would be two spaces or two times. It would only follow that there would be a physical space and a physical time, which would be experienceable in some sense of that term, but the experiencing of which would itself have been made possible by the application of the Forms of Plurality in the first place. In the end, there may be some very interesting application of the Forms of Plurality to the world of the physicist: for instance, in the interrelationship of physical space and time which may be an analogue to the conceptual interrelationship of which more will be spoken in the sequel. It should not be surprising that the physical world is to some extent a mirror of the mental world, or, that the subject knower is a microcosm of a macrocosm. This has been touched upon earlier as it is impossible to carry out the work of epistemology without a necessary minimum of reference to ontology. But this discussion may be deferred. It is important, at the outset, only to allay the fears of the reader that one may, as Kant - despite his denials - create two spaces and two times, and so solidify the popular impression that philosophy has little application to the world in which one lives. In the case of Kant, for example, the space and time discussed in his First Antinomy appears, on the face of it, different from the space and time that he discusses in his Transcendental Aesthetic.

The space and time of the person in the street or the physicist is real. The capacity for pluralization is the condition for being able to represent the space and time of the person in the street or the physicist in the first place. But, the capacity for pluralization is in no way responsible for creating physical space or physical time. Questions concerning the reality of, or the nature of physical space or time, are not dependent upon the investigation of the Forms of Pluralization. No claim is being made herein that space or time are *only* mental forms. At the moment nothing at all is being said concerning the reality status of space and time, *vis-à-vis* the world of the physicist. For the moment, at least, it is better to keep these realms distinct. It has been the unfortunate aspect of the philosophy of Kant that he did not sufficiently clarify the distinction between physical space and his own concept of space to the extent of denying the possibility of physical space altogether. But, such a denial is not necessary to the construction of a philosophy in which space and time are understood to be Forms of the representation of Plurality. And, the construction of a philosophy in which space and time are Forms of the representation of Plurality is a beginning step in the direction of understanding philosophical concepts as unique to the discipline of philosophy.

Without further ado, the task of defining the Forms of space and time must be begun. At the outset, it was stated that space and time were interdefinable. Space cannot be defined without reference to time and vice-versa. It is now the time to make this definition more formal and specific. *Space may be defined as the transcendental condition (in the Kantian sense) for the possibility of the representation of multiple existence at the same time.* Space cannot be defined without reference to its sister existence, time. In order for two things to be represented as positioned next to each other, or in some spatial relationship, they must be represented as existing at the same time. If two things were represented as existing at different times, they could not be represented as being positioned in some spatial relationship to each other.

Kant approached the proper definition of space when he stated that only on the presupposition of time can one represent to oneself a number of things existing at one and the same time.[2] Unfortunately, Kant thought that time alone was being presupposed, which it surely was not, as will be evidenced by the arguments below. Were Kant to have considered the inter-definability of space and time, his entire argument would have been clearer and more cogent.

Time may be defined as the transcendental condition (in the Kantian sense) for the possibility of the representation of multiple existence in the same space. Time cannot be defined without referring to its sister existence, space.[3] Even Aristotle's classical definition of time as the 'Number of Motion' makes a hidden reference to space, since motion, whether circular or local, must be through space. Aristotle's notion is demonstrated by every kind of analogue timepiece from the sundial and hourglass to the analogue clock. Time cannot be demonstrated without showing a movement through physical space. Every attempt to image time involves a movement or a suggested movement through space such as is signified by the phrases, 'Time's Arrow' or the 'River of Time', or 'Water over the Dam', or 'Water under the Bridge' or, of course, 'Time Flies'. In the case of the digital watch or clock, time is already included symbolically by the use of numbers, which indicate greater or lesser. Greater or lesser cannot be understood save by a reference to degrees of extension or numerical representations. Digital representation is thus derivative from a more primitive spatial representation and is ultimately dependent upon this spatial representation for its intelligibility.[4] It could be said that time is also indicated literally since each number must take up a certain amount of time in its appearance and a certain amount of space in its representation. But, this consumption of space and time is in terms of the use of space or time in the representation of the numbers, and not in terms of the representation of time proper.

It must be borne in mind that one is not now referring to physical space or time, although there may be some analogical empirical overlap. That to which one is referring is the one and only one Form of Plurality that is space, and the one and only one Form of Plurality, that is time. It is a totally separate issue as to whether there are multiple physical spaces and times or not. The point is that there can be only two forms of non-conceptual plurality. It makes no sense to refer to further distinct forms of plurality, as it would not be clear how these could be distinct from one another unless there were another form of plurality and so on *ad infinitem*.

It is interesting to note that in some key respects physical space and time share characteristics with the Forms of Plurality. But, one must not for that reason make the mistake of identifying the two *simpliciter*. Such an identification will make philosophy appear, in the end, to be a competitor with physics. While there is a definite and highly interesting connection between the philosophical use of the terms 'space' and 'time' and the empirical use to be specified in the sequel, they are by no means to be identified.

As Forms of Plurality, it has been noted that space and time cannot be utilized in isolation from each other. Space *and* time are dual transcendental conditions of the representation of multiple existence. This connection between space and time does not appear in the philosophy of Kant. Earlier, space was defined as the transcendental condition for the possibility of the representation of the existence of more than one in the same time. There is no way in which space can be defined by itself. The simple externality which Hegel thought sufficed as a definition of space is already parasitic upon the definition of time, as what is represented as outside of or external to another, is and can be represented as outside another, only at the same time. Otherwise, its outsideness to a specified other would not be discoverable. Time is required in order to complete the definition of space. Similarly, in the definition of time, space is required for its complete exposition. It is impossible to indicate time without recourse to a spatial metaphor. If there is no specified location in space, two times cannot be distinguished from each other. Even in memory, one time is distanced from another time only by visualizing two separate locations in which the memories can exist or to which the memories refer.

There are two and only two modes of representing a plurality. These two modes are space and time. Things can be represented as distinct from one another by reference to either space or time. Space and time in themselves, however, can only be represented by reference to each other. Space and time are mutually implicated in each other's definitions. They are both equally primitive terms. But, they are also incomplete. They complete each other when they are added to each other. In terms of their use, however, they need not, although they often do, make explicit reference to each other. (They must of course always imply each other implicitly.) In terms of their definition, they are inextricably intertwined with one another. While Kant did not note the necessary connection between space and time, he did isolate space and time from other concepts and in the first *Critique* he recognized them as constituting two sources of knowledge (B55).

Time

Time is the transcendental condition for the possibility of the representation of multiple existence in the same space. There is no other way to indicate time without implicating at once its equally primitive sister concept of space. In Kant's terms, this would be considered 'succession', which is, for Kant, only one of the three modes of time.[5] It is, however, in the present author's view, the defining characteristic of time. Succession enables one to represent two objects as occupying one and the same space – namely, at successive and thus different times. Succession is not a mere mode of time; it is at once also a reference to space. One cannot represent succession without referring to space. One cannot represent how things can at once be and not be, unless they are and are not in something, or in relation to something, which remains constant. The constant frame of reference for non-existing and then existing times is space. There is no other way of representing succession, whether one is referring to a succession of visual images, scents or musical notes. When one holds musical notes in mind and is aware that one note preceded another, the succession of notes can be represented as preceding and succeeding only by a precession and a succession *within* the same consciousness which is represented as a common mental space.

Succession is nothing other than the nature of time itself. Kant considered that there were three modes of time, but of these three, succession (from the point of view of the present work) is the most fundamental, since it itself is the way in which pluralities can be represented or distinguished in the same space. The mode of simultaneity, for example, is not essentially a mode of time at all but is the form of space masquerading as time. Simultaneity - or two things happening at the same time – is really a definition of space. There is no way in which one can represent multiples of time without a reference to space. If one attempts to represent a multiple existence at the same time, one must employ the form of space. Two things can be represented as occupying one and the same space only by being represented as existing at different times. When two things are spoken of as happening at the same time, they cannot be represented as being distinct from each other unless a spatial representation is employed. Otherwise, they would not be experienceable as *two* in the first place. If truly simultaneous, there would only be one. There is no way of representing simultaneity (which involves two) without invoking the concept of space.

The other mode of time of which Kant speaks is that of duration. Duration (from the point of view of the present work), strictly speaking, is not a mode of time or of space. It is a species of successions held together by memory. But it, too, involves a reference to space. The constancy given by duration is borrowed from space. If something is spoken of as lasting a long or a short time, it has already been represented with a spatial metaphor.

Kant was not quite correct to conceive of these three characteristics as modes of time. They are one and all references to space as well. But, of the three, the most truly representative of time is that of succession. Duration is really a species of succession as it involves a multiple of successions, which are based in part on memory. Simultaneity appears to make a clearer inroad into the concept of time in that it refers to the present, but is not uniquely characteristic of time, since it cannot distinguish itself sufficiently from space. In fact, simultaneity is another way of defining space itself. The definition, 'two things represented as happening at the same time' is a definition of space while employing the concept of time. The element of plurality is contributed by space; the element of constancy is contributed by time. In simultaneity, the aspect of plurality is contributed by space. In duration, the aspect of plurality is contributed by succession. Therefore, duration is derivative from succession. Succession, or one event replacing another event, is the condition for duration.

Succession is the best or, strictly speaking, the only definition of time proper. It is the best or the only definition of time in that it explains the possibility of the representation of plural existences as existing at the identical spatial locus. Plural existence at the identical spatial locus can only be represented through the phenomenon of existential replacement. Succession is the existence of one after another, or the very mode of existential replacement itself. This remains true whether the space to which one has reference is logical space or the space of the imagination. In the concept of simultaneity, the necessary existence of plurality is provided by space. In order to represent two things as existing at the same time, the twoness is contributed by space. In succession, the existence of plurality is contributed by time. In the case of succession, space is the constant and time is the mode that contributes the sense of plurality. Duration is not a species of time at all, but involves a memory of successions. Succession is another word for time itself. Succession is the only modality in which a multiple existence can be represented within the same space. Space is a limited form of plurality, since it cannot account for a plural existence inside of itself (this is the truth contained in the Kantian insight that a multiple of space is itself another part of the selfsame space). In order to account for the representation of a more than oneness that can exist in the same space, space requires the concept of time. The only way in which one can represent plural existence in one and the same space (precise location) is by succession, which is therefore the defining characteristic of time itself.

Kant should not have wished to affirm that physical space and time did not exist, since to argue that space does not exist physically is to make a claim about the thing-in-itself about which, *ex hypothesi* Kant has nothing to say - or at least, according to his precepts is supposed to, contrary to his practice, say nothing.[6] Kant's practise, however, at times, violates his principles. For example, there are moments during the first *Critique* when Kant speaks about the thing-in-itself as a cause, but strictly speaking, such statements on his part are inadmissible. At other points during the first *Critique*, he says that all that one can say about the thing-in-itself is that it exists. For Kant, indeed, the thing-in-itself must exist. However, it is difficult to comprehend, on his terms, how he can say even this. Though, of course, if he were to say that the *ding-an-sich* did not exist, then all appearances would be purely subjective, and he would be reduced to solipsism. It is not clear that one can escape solipsism simply on the basis of an assertion which is not allowed on the basis of one's philosophical principles, but many liberties are allowed in the case of great philosophers.

What has been offered in this present account is the phenomenological characterization of space and time. There is no ontological claim advanced as to the existence or non-existence of physical space and time. From the strictly phenomenological point of view, the existence or non-existence of space and time is not contemplated. Kant considered it important to allege that physical space and time did not exist. In the treatment of space that occupies this present work, the above difficulty is well avoided. In this present work, it is not necessary to argue that there is one and only one space. In fact, from the standpoint of this present work, it is not appropriate to apply the concept of plurality, which depends upon space and time, to space or time. From the standpoint of this present work, it makes little sense to say anything about the quantity or the extent of phenomenological space or time (whether space and/or time are one or many, finite or infinite), since all quantitative representation requires spatial and temporal representation. One must employ the concepts of space and time proper whenever one phenomenologically imagines either an infinite divisibility or extensibility of space or time.

From another point of view, since this work is concerned (at this stage), only with the world of appearances, and not with things-in-themselves, the question of the existence or nonexistence of physical space, or the relationship of physical space to phenomenological space falls outside of its scope of inquiry. Or, to put this in another way, whether there is more than one space - in addition to the form of representation - is at this moment - of no interest to the phenomenologist, and thus does not affect or invalidate any characteristics of phenomenological space. In short, it is not necessary to argue that there is one and only one space, and it is not really possible to raise the question of the plurality of space, since the concept of space is always presupposed in any notion of plurality.

What one could say - and what Kant perhaps should have said - is that the same Form of Plurality that is space is presupposed in any and every plurality including the possible pluralities of physical spaces. One reason Kant did not say this is because he did not want to talk about physical space (which implies the existence of space outside of the mind of the subject knower) at all. But, this is of no fear to the phenomenologist whose claim is not that one cannot talk about physical space outside of consciousness, but that anything that one says about it will lack universality and necessity, and thus will not be of first order philosophical interest. One may speak about time and space outside of consciousness as a metaphysician, but this first depends upon one's metaphysics receiving some phenomenological grounding.

Kant's Mistake

Kant's Copernican revolution was based on his premise that '... we can know *a priori* of things only what we ourselves put into them' (Bxviii). While such a point of view appears to be a possible explanation for the origin of the *a priori*, it is not obvious that it is a sufficient explanation. What gives the finite human mind the power to formulate what is *a priori* and give *a priori* laws to the universe? Kant's explanation presupposes a relationship between the mind and the universe which requires a prior bond which he does not provide. Why does the human mind impose its forms on the universe? How is this possible? Why should the human mind impose its forms on the universe? Why do the forms of the human mind fit the universe? Kant's explanation makes it appear that the universe depends upon the human mind for its functioning. A series of questions follow in its wake. Do other animals also impose the forms of space and time on the universe? Is it possible that Kant has confused the phenomenological experience of apodeicticity with a putative explanation for its possibility?

Kant's proofs of his Copernican hypothesis '... from the nature of our representations of space and time ...' (Bxxiia) assume the form that for human beings representation itself is impossible without space and time and therefore space and time are forms of human representation. While this could explain why space and time were always employed in the representation of any appearance whatsoever, it is an insufficient explanation to account for how it is possible to experience the strict universality and apodeicticity of discovered truths about space or time.

The phenomenological fact that Kant has discovered is the appearance of strict universality and apodeicticity. Kant may have inferred (rightly or wrongly) that the basis or the reason for why phenomenological certainty and universality exist is that the sources of such certainty and universality are given by the subject knower. This inference is similar to Vico's principle that one only knows for certain that which one has created. But Kant's explanation that the unavoidability of the employment of space and time arises from the fact that they are forms contributed by the subject knower is not the only explanation that would account for the unavoidability of the employment of the forms of space and time. What is more, without a further account of the subject knower and the relationship of the subject knower to the world, an inference to the possible origin of the forms of space and time is not enough to account for the fact that truths about space and time both apply to the world and are capable of being experienced in their apodeicticity. While Kant could argue that he may not be accused of applying the forms of space and time to the world, the Copernican analogy is of his own invention. In any case, Kant cannot escape the question of how the subject knower can experience apodeictic knowledge.

From the point of view of this present work, apodeicticity and universality are qualities of experience. Perhaps, the reason why Kant shied away from attributing necessity and universality to experience was that he was overly influenced by Hume's account of experience as empirical. In mathematical truth, the experience of 'can never be otherwiseness' cannot be given by empirical experience. And yet, it is this very 'can never be otherwiseness' which is experienced and therefore known as true that imparts to mathematical truth its unique form of necessity and universality. If this 'can never be otherwiseness' could not be experienced, there would be no detectable difference between mathematical and empirical truths, which is surely contrary to fact, and Kant's own claims to the synthetic *a priori* character of mathematics would be invalidated as well.

In Kant's famous opening sentence to the second paragraph of the Introduction to his first *Critique*, he states that all knowledge begins with experience but it does not all arise out of experience. But this first sentence of his second paragraph is a veritable Kantian Pandora's box. It must be noted that *a priori* knowledge is a species of knowledge and certainly is subsumed under the heading of 'all knowledge'. But one must consider what characteristics belong to *a priori* knowledge. Kant states that, 'Necessity and strict universality are thus sure criteria of *a priori* knowledge, and are inseparable from each other' (B4). Could it be then for Kant that one is directly aware of *a priori* knowledge in experience? One can recall that for Kant, '... geometric propositions are one and all apodeictic, that is, are bound up with consciousness of their necessity ... ' (B41). It must be borne in mind that Kant states that one can become conscious of the necessity of judgements that (for Kant, given Kant's account of experience) could never have been learned from experience (B12). Kant thinks that one can become conscious of necessity. Consciousness would certainly appear to be a species of experience. The subject knower thus seems to possess knowledge that experience, defined by Kant as empirical knowledge, does not warrant. For Kant, '... empirical knowledge is experience' (B166). And empirical knowledge (experience) cannot yield necessity and strict universality (B5). Then, how can Kant claim that we do possess synthetic *a priori* knowledge? Is it purely as a transcendental condition? But some of Kant's examples of synthetic *a priori* knowledge are not examples of knowledge only as a transcendental condition. The problem may be that experience is not exhaustively defined by empirical knowledge.

Perhaps, Kant is himself too preoccupied with the transcendental condition for the possibility of x form of argument to take sufficient notice of what he has uncovered in his famous phrase. Whether or not the experience of the necessity or the universality of the truth in question is actually derived from the putative source of the truth is a different question. Indeed, from whence the necessity or the universality of the truth arises is a totally separate issue from the fact that such necessity and universality must nevertheless be experienced. Regardless of what one considers the source of such necessity and universality, such necessity and the universality must form part of the experience, which is an astonishing fact to be explained and one that Kant does not explain. It is of interest to further explore Kant's famous phrase that heads the beginning of his second paragraph of the B Introduction to the first *Critique*. In Kemp Smith's translation of Kant's frequently quoted sentence 'But though all our knowledge begins with experience, it does not follow that it all arises out of experience', experience appears to be a necessary, but not a sufficient condition for knowledge. What is left unexplored in such a statement, however, is, how is it possible that if experience were (for Kant) inadequate to account for knowledge, that knowledge can nonetheless arise within it? This is Descartes' question, but one which Kant does not address.

For Descartes, the question assumes the form of his causal proof of the existence of G-d. Since humans are finite and thus (for Descartes) not capable of generating the idea of perfection and nevertheless (according to Descartes) the subject knower does possess such an idea, this idea must come from G-d (as there cannot be more in the effect than there is in the cause), and therefore G-d exists. To rephrase Descartes' point, since experience does not allow the subject knower to possess knowledge, and such knowledge is possessed, such knowledge must come to the subject knower from a higher source of reality. Even this rephrasing of Descartes does not quite solve the problem since it still remains a mystery (from the point of view of the present work) how G-d could cause such an idea in human beings without in some sense deifying human beings. Otherwise, it is not clear how finite beings could possess the idea of G-d, since by definition the finite cannot encompass the infinite. On the other hand, if the human being could be deified, then it would no longer be clear in what sense the human being would remain finite. It could be argued that the human being possessed both a finite and an infinite part, but it is difficult how there could be any communication or interaction between the finite and the infinite parts and how one could justify that the two parts belonged to the same entity.

An interpolation of Spinoza's thought presents a possible solution to Descartes' problem, but it is not itself problem free. Man's knowledge of G-d's perfection is nothing other than G-d's self-knowledge and hence there is no paradox of the finite knowing the infinite. It is in so far as man is Divine that man knows perfection. One can carry this argument one step further. One can argue that Descartes' causal proof of the existence of G-d was the Deity's proof of Divine existence, or in other words, the argument of the Divine *Cogito*. For the Divine, the argument would assume this form. There cannot be more in the effect than there is in the cause. I (G-d) have the idea of perfection. An idea of perfection can only be generated by a perfect being. Therefore, if I have such an idea, I must exist in order to have caused such an idea. From the point of view of man, if one were to be capable of comprehending perfection, that comprehension would be the very moment of Divine self-comprehension and a proof at the same time of Divine existence. Of course, the problem pointed to above of the finite and the infinite remains which seems to be a problem that even Spinoza cannot entirely overcome. Spinoza attempted to solve this problem via the immanence of the infinite in man for it is no longer in the capacity of man being finite that she or he understands perfection but only *qua* her or his being G-d.[7] Contrariwise, one could render the entire problem nugatory by denying that one can comprehend the concept of perfection. If perfection is indeed an incomprehensible concept, then Descartes' argument (and Anselm's) cannot be made. Indeed, if perfection is incomprehensible even to the Divine, then the argument of the Divine *Cogito* fails as well.

If the above formulation of the problem were to be applied to Kant, it would be clear that if within experience, knowledge were to be possible (regardless of whether or not the knowledge originates from inside or outside of that experience), it nonetheless must be possible to experience that knowledge within experience. *One cannot confuse the source or the origin of the knowledge with the fact of the arising of and the awareness of the knowledge during experience.* It is irrelevant if the condition for the possibility of knowledge is that the subject knower contributes its form (the counterpart of the Cartesian office of G-d). The fact would remain that the knowledge would and must be experienced, nevertheless. Kant's problem of how knowledge is possible during experience is no more solved by claiming an internal origin of necessity than Descartes' problem of how knowledge (of perfection) is possible during experience is solved by claiming that it is bestowed by G-d (an external origin).

Space

Space is the transcendental condition for the possibility of representing the existence of multiplicity during one and same time. While space is also required for representing a length of time, it is more fundamentally defined as the transcendental condition for the possibility of the representation of a multiple existence at one and the same time. Time shows its incompleteness in that it cannot provide a pluralizing mode within itself. If time were divisible inside of itself, it would still be part of one and the same time, and hence could not represent a distinction within itself, except through recourse to the representation of space.

Two objects can be represented as existent at the same time by their representation as alongside of each other. Two objects can only be represented as existent in the same space by their representation as occurring in succession. These are the two most fundamental modes of representing pluralities and all forms of the representation of multiplicity are forms of one or the other of these most fundamental forms.

Space is inherently incomplete as a form of pluralization because it requires the employment of the constancy of time for its complete definition. Space is the transcendental condition for the possibility that more than one object can be represented in the same time. Space is inherently incomplete without its twin sister of time. For example, if one were to assume Hegel's definition of space as a form of externality or outsideness, it would be impossible to represent the existence of external multiplicity at different times. Simple externality will not suffice as a definition of space, since it cannot be guaranteed that a multiple of existences can be represented as existing in one and the same space, if the representation is taking place at two different times.

Time is inherently incomplete as a definition, because it depends upon the constancy of space for its complete definition. Time is the transcendental condition for the possibility of the representation of multiple existence in the same space. The existence of simple succession without reference to space would permit no knowledge that the plurality was a plurality of one and the same existent. One moment of time can be represented as replacing another moment of time, only within the same space.

Space and Time Revisited

Space and time are both primitive terms. There is no other more primitive term or terms by which space and time can be defined. Taken together, they make up a complete repertory of the modes of pluralizing. It is only possible to possess one mode of plurality because of the existence of the other mode. They do not merely complement each other; they help to constitute one another. Objects can be represented as alongside each other in space only by the representation of the objects as existing within the same framework of time. In different frameworks of time, objects cannot be coherently represented as alongside of each other in space. One can represent objects as before or after one another only by representing the objects as existing within the same space. Objects cannot be coherently represented as before or after each other if they are represented as existing in different spaces. This is the case whether one refers to physical space or the space of the imagination.

Alongsideness and succession are the only two modes of pluralizing. All other modes of pluralizing have been shown to be reducible to either alongsideness or succession. Duration, for example, is a sequence of successions. Simultaneity is dependent upon space and in reality is not a mode of time at all. To represent simultaneity, two things must be represented as occurring together, in short, by utilizing the framework of alongsideness, or, space. In order to represent the simultaneous occurrence of two things far apart from each other geographically, a spatial bridge between the two events must be imagined. Such a mental bridge is not, as Kant thought, the unity of consciousness, because the cognition of simultaneity is not comprehended by the abstract and incomplete notion of unity. That there must be a unity of consciousness in which two representations inhere is not (or, should not have been) what Kant meant by the transcendental unity of apperception, for such a meaning would constitute a paralogism in the Kantian sense. It would be to treat the Self as a substance. This meaning of unity, in any case, would be to posit a mind outside of phenomenological consciousness in the terms of the present work - the other side of Husserl's mistake of positing an object for consciousness. This meaning of the transcendental unity of apperception would be the positing of consciousness for an object. All that appears as consciousness, the domain of the phenomenologist, is framed by the outlines of space and time. In any event, by the concept of the transcendental unity of apperception, what Kant must have or should have meant was that there must be a transcendental unity for consciousness. This transcendental unity is part of what is provided in the present argument by space for time and by time for space. Simultaneity can only be represented by the form of togetherness or alongsideness, which is space. But, space itself is only possible as a mode of pluralizing under the condition of the possibility of the employment of the form of time.

All of the foregoing is just another way of saying that space and time are interdependent. They cannot be defined apart from each other. Duration is really a string of successions; simultaneity is really another way of referring to space. Strictly speaking, simultaneity is not a mode of plurality at all; it is a mode of unity. But, in unity there is no simultaneity, as simultaneity always refers to two. Simultaneity is, when closely examined, a spurious concept. There is no way in which two things can be represented as existing at the same time unless one makes primary use of the concept of space. In the attempt to state this, in effect, one states the definition of space and not a description of a mode of time. One inadvertently elicits the primitive term of time, but only insofar as it is involved in the definition of space. Simultaneity does not refer to the essential nature of time itself; it is the way in which time expresses itself as space. Space is the necessary condition for the possibility of being able to represent two events as occurring at the same time. Time cannot distinguish within itself temporally. Internal time consciousness, or duration, is space. In order to notice that two things can take place at the same time, it is necessary to employ the form of space.

Succession is the temporal form of the appearance of plurality. In order to represent succession, one must refer to the fact that there are two, in one and the same space. This is the essential feature of time, but it cannot be fully expressed in the absence of the employment of the constant framework of space. Succession is time proper; alongsideness is space proper. Together, succession and alongsideness fully define each other.

Space and time are Forms of Plurality rather than concepts of plurality, because they are not used to recognize plurality in the same sense that one would employ the empirical concept of a porcupine as a rule for the recognition of an empirical example of a porcupine. Space and time are necessary conditions for the concept formation of the concept of plurality as such in the first place. One could not understand conceptually what could be meant by 'more than one' unless one possessed reference to the concept of space-time. What is especially interesting about all of this is not only that space and time are required as a condition for the representation of plurality in the first place, but that space and time are required as conditions for the possibility of their own self-representation. Equiprimitive, they constitute the correlative defining characteristics of each other. One cannot represent space without time and one cannot represent time without space. It is not simply or solely that they generate each other; they do not make sense as concepts without each other. It may be left either to cosmologists or to a later generation of philosophers practising well-grounded metaphysics to posit the possible consequences for world making. As a pure phenomenologist, one is only interested in showing the logical interconnectedness of the concepts of space and time. It is of great interest that they are logically incomplete without each other. For phenomenology proper all that can be hoped for is to show how the concepts themselves only make sense when considered together. This discovery in and of itself may be of some use in paving the way to other discoveries.

When the philosopher ventures outside of phenomenology proper, the phenomenological discovery of the interrelatedness of space and time may be a rich field for analogical conjectures concerning reality. But, such conjectures, as fascinating and as fruitful as they may be, can never possess the certainty of truth that belongs to the discoveries of phenomenology proper. They can, however, make claim to a truth status higher than that which belongs to pure speculation, because they are grounded in previously discovered phenomenological truths. Such conjectures belong to the special theory of the application of metaphysical truth.

Kant asserts that the thing-in-itself exists even though he does not directly attempt to prove the existence of the thing-in-itself in the first *Critique*. His Refutation of Idealism and his Second Analogy are implicit attempts to prove the existence of the thing-in-itself. The entire Refutation of Idealism is based upon the unalterable order of experience, which itself is based upon the experience not of appearances, but of external reality. While this is one of the few places in the first *Critique*, where Kant speaks in such a non-Kantian fashion, how else is he to counteract the possibility that his appearances are all illusions, unless the thing-in-itself is behind all appearances? It is true that he would prefer that the thing-in-itself were simply the concept of the appearance when it is not appearing, but the Refutation of Idealism and the Second Analogy demand more than this purely negative or limiting concept of the thing-in-itself. Indeed, in his Refutation of Idealism, one can find Kant making such explicit statements with his own emphasis as, 'I am conscious of my own existence as determined in time. All determination of time presupposes something permanent in perception. This perception of this permanent is possible only through a thing outside me; and not through the mere representation of a thing outside me; and consequently the determination of my existence in time is possible only through the existence of actual things which I perceive outside me' (B276).

In the Second Analogy, despite the intriguing defenses of Kant that have been offered by his commentators, in the opinion of this present, respective author, when Kant freely speaks of the perception of a ship going downstream in order to show how the order of succession is dictated by an external order, he thereby implies that he has access to an externally derived order or an order caused by a thing-in-itself.[8] But such access is not permissible according to the essential thesis of the first *Critique* that all knowledge is only of appearances, not of things in themselves. Similarly, Kant's Refutation of Idealism is unsuccessful, because if it were successful, it would undermine this essential thesis of the *Critique*.[9]

While this is not the place to offer a complete commentary on Kant, it may be said in passing that the Second Analogy can be better understood if it is understood not as an argument which attempts to distinguish between an objective and a subjective succession or a public and a private succession, but rather is understood as an attempt to define succession itself (whether real or illusory, objective or subjective, public or private).[10] Succession, as opposed to simultaneity, requires a connection of the 'after' to the 'before'. Thus, the Second Analogy is actually a definition of time. Whether this time is the time of a dream or the time of reality is irrelevant. Kant's argument that the order of succession in watching the ship is irreversible is just as true of the dream order as of any other (unless one is in the state of lucid dreaming and can also re-arrange the sequence of events occurring within the dream while staying within the dream state).[11] Kant contrasts the ship to the house in that he thinks that in the case of the house one can reverse the order of succession of perception. But, this comparison is infelicitous. Where one starts (top or bottom) is immaterial to the fact that once one starts, the order cannot be reversed. Or, if the order were reversed, it then would make up a different ordering sequence (bottom, top, middle, or whatever). That Kant thinks in the case of the ship, one cannot reverse the order is only because Kant is assuming (contrary to what he later states) that he could tell at this moment that he was not dreaming. With the exception of the state of lucid dreaming alluded to above, in which one can become aware of the fact that one is dreaming while one is dreaming without awakening from the dream and furthermore, possess the skill of re-ordering the succession of events in the dream, even if Kant were to be dreaming, the ship would follow the same laws as the house. Thus, no particular example of seeming irreversibility can demonstrate objective succession or even function as a successful criterion of demarcation of reality from illusion. It can only demonstrate the phenomenological reality of time as opposed to space.

Space, Time and Metaphysics

With respect to the category of propositions about phenomenological experience, metaphysical propositions may be classified under the synthetic *a priori*. While much attention has been devoted to either maintaining that only synthetic *a posteriori* or analytic *a priori* propositions are valid (Ayer) or that the synthetic-analytic distinction is not a valid one (Quine, 'Two Dogmas of Empiricism'), comparatively little attention has been paid to the synthetic *a priori* itself.

Metaphysical propositions can be classified under the category of the synthetic *a priori* in the sense that they are necessarily true (or false), and are not mere conventions. Metaphysical truths are discovered rather than stipulated, and thus cannot be otherwise. Metaphysical truths are recognized and certified within phenomenological experience - although their 'truthafiability' or 'truthalizability', to coin some terms, does not originate from phenomenological experience. (Otherwise the entire argument of this present work would be circular). This must be the case, as has been argued above and as will be returned to in the sequel, since phenomenological truth will possess a metaphysical correlate, which must precede and make the phenomenological truth possible. Such considerations form part of the general theory of the application of metaphysical truth.

One may begin with the example of time being defined as the transcendental condition for the possibility of the representation of the existence of more than one in the same space. While Hegel defined space as otherness to itself (pure externality), he was not aware or at least not obviously aware that this required time as well - for pure externality means otherness to itself at the same time.[12] Kant, in his definitions of the three modes of time (duration, simultaneity and succession) did not consider that all three of these *modi* required the employment of the concept of space in order that their definitions be complete.

Since neither Kant nor Hegel conceived of the interdefinability of space and time, the problem of whether space and time could be defined without each other was not taken under consideration. Space and time were simply defined without each other, without the realization that this itself required a justification. While ostensibly, space and time have historically been defined in independence of each other, in reality, of course the definitions have always been parasitic upon each other. As mentioned earlier, even Aristotle's famous definition of time as the number of motion tacitly makes use of a spatial metaphor in that space is required for the numbering of time whether that space be digital or analogical.

The definition of time is a metaphysical proposition, because it is necessary and discoverable in pure phenomenological experience and cannot be otherwise. In some usages, the term 'metaphysics' and the phrase 'synthetic *a priori*' are interchangeable. However, there are usages in which the term 'metaphysics' is not interchangeable with the phrase 'synthetic *a priori*'. This will receive further qualification below.

Science and Metaphysics

According to the famous statement of Minkowski, space and time are related in a space-time continuum. This fact should not be surprising, since according to the analogies of experience to be discussed in Chapter Seven, the microcosm is always a microcosm of the macrocosm, such that the macrocosm is always contained in the microcosm. The phenomenological experience both points to and is a reflection of metaphysical reality. Metaphysical reality cannot be spoken of directly, but only as an analogy of experience, so that metaphysical *reality* statements are not strictly speaking synthetic *a priori* statements. This is the truth contained in Kant's denial of metaphysics.

Science and metaphysics both require entirely different criteria of truth: one cannot apply the standards of scientific truth to metaphysics and those of metaphysics to science. But there is only one truth (only one macrocosm). Consequently, science and metaphysics will converge in their truth statements, while the criteria applied to each discipline in order to ascertain what respectively is a metaphysical truth and what is a scientific truth will be radically different. The criteria of scientific and metaphysical truth are different, but their ultimate truth content can be connected via an analogical bridge.

Scientific facts may not be utilized as a way of establishing metaphysical truths or vice-versa, but it nonetheless remains the case that the two truths must relate to each other in the ultimate realm of truth. Scientific and metaphysical truths can be discovered and in fact must be discovered to be true independently of each other. But once discovered, the truth of one may be used as a clue to the potential, as yet to be discovered truth, of the other. In this way science and metaphysics can cooperate as two pathways which may serve as guides to each other in the manner of two partners collaborating in a search for ultimate truth. From afar they may applaud the other when their alter ego discovers the selfsame truth that they themselves have discovered in the course of their own intrinsic inquiries.

Notes

1. *Parmenides*, 146D. *Cf.*, Cornford's analysis of this passage in his great commentary, *Plato and Parmenides*: 'The difference here defined is, as we have remarked, numerical difference, or at least primarily that, though the argument would apply equally to conceptual difference.' *Cf.*, F. M.

Cornford, *Plato and Parmenides*, London: Routledge & Kegan Paul, 1939, p. 159.

2. 'Only on the presupposition of time can we represent to ourselves a number of things as existing at one and the same time (simultaneously) or at different times (successively)', *Critique of Pure Reason*, B46 (Kemp Smith translation).
3. For Philo Judaeus, 'the extension of Heavenly motion is time'. *De opificio mundi*, 26. Chinese philosophy recognized the inter-ontological nature of space and time even earlier than Philo. In ancient Chinese philosophy, space and time are referred to as twin concepts. What is characteristic of the present, respective work is that a philosophical argument (and not an argument either from speculative metaphysics or empirical physics) is offered for the inter-ontological nature of space and time. In his two volumed, *Space, Time and Deity*, Alexander does take the position that space and time have no existence apart from each other, but his argument appears to require metaphysical postulates and is obscure. Eugene T. Gendlin employs ingenious arguments and illustrations of his own device and arguments and illustrations similar to those of the present, respective author to demonstrate that Kant relies upon space in order to define time. *Cf.*, 'Time's Dependence on Space: Kant's Statements and their Misconstrual by Heidegger', in Thomas M. Seebohm and Joseph J. Kockelmans, (eds.), *Kant and Phenomenology*, Washington, D. C.: Center for Advanced Research in Phenomenology and University Press of America, 1984. For example, Gendlin points out that when Kant uses the term 'figurative', he means 'spatial' (p. 152). For the present, respective author, what Gendlin demonstrates is that time cannot be defined without employing space. In this connection it is worth noting Copleston's point in his volume on Kant in his *History of Philosophy* that when Kant referred to Euclid's definition of a straight line as the shortest distance between two points that Kant emphasizes that quality (straight) is employed to define quantity (shortest). But the more relevant point would be that the quality of straightness can only be shown through the use of space not that it is a quality rather than a quantity. Despite showing his awareness of the dependence of time on space, Kant never does define time in terms of space or vice-versa. In his *Inaugural Dissertation*, Kant does seem to approach the concept that space cannot be defined without reference to time. *Cf.*, *Kant's Inaugural Dissertation and Early Writings on Space*, John Handyside, (trans.), Westport: Hyperion Press, Inc., 1988, p. 65.
4. While it has been advanced that geometry is based upon space and arithmetic on time, it is not quite correct to say that arithmetic is based on time. Arithmetic is ultimately based upon space as well since the basic operation of incremental increase or decrease can only be understood by reference to space. That it was thought that arithmetic was based on time was a result of considering that time was required in order to convey the concept of incremental increase or addition without an alteration in shape. However, that duration (not time *simpliciter*) was required in order to

perform the operations of arithmetic is only incidental since duration is also required in order to carry out geometrical proofs. What is essential to the understanding of arithmetic is the understanding of increase and decrease and such understanding requires space proper and not time. In an analogous fashion it can be shown that space and not time is what is required for the proper understanding of algebra, the basic operation of which is transposition and the basic relationship of which is identity and difference. What is correct, and what may have been the misleading factor in the identification of arithmetic with time is that the ordinal numbers are indeed only comprehensible under the framework of time. Without time proper, e.g., succession, the concept of an ordinal number could not arise. It is a separate question as to which operation is more basic, numbering or counting. What is important to realize is that both operations depend upon plurality which is more basic and which manifests itself either spatially or temporally. Number is a labeling or articulation of space-time. Husserl saw that cardinal numbers arise because of experienced pluralities.

5. *Critique of Pure Reason*, B219.
6. While even in the A edition, Kant contradicts himself on the question of whether the thing-in-itself must exist or if it is only a limiting concept, he does make clearer statements on behalf of its existence: 'Unless, therefore, we are to move constantly in a circle, *the word appearance must be recognized as already indicating a relation to something*, the immediate representation of which is, indeed, sensible, but which, even apart from the constitution of our sensibility (upon which the form of our intuition is grounded), *must be something in itself, that is, an object independent of sensibility*.' *Op. cit*., A252 (Emphasis added). He also refers to the thing in itself as the true correlate of sensibility which of course it must be if one is to avoid subjective idealism. *Cf.*, B45. In a revealing passage in B72, Kant states that, 'Our mode of intuition is dependent upon the existence of the object, and is therefore possible only if the subject's faculty of representation is affected by that object'.
7. For the great Spinoza commentator, Wolfson, while the human mind is part of the infinite intellect of G-d, it seems that it is not a part of G-d's essence itself. *Cf.*, Harry Austryn Wolfson, *The Philosophy of Spinoza*, Volume II, Cambridge: Harvard University Press, 1934, p. 50. Spinoza says, ' that our mind, in so far as it truly perceives things, is part of the infinite intellect of G-d ... ' *Ethics*, II, Prop. XI, Corol. In so far as the human mind is part of the infinite intellect of G-d, according to Wolfson, it gets its ideas from G-d (*Ibid.*, p. 157). Thus, the argument offered above that G-d is fashioning the causal/ontological proof is not without warrant.
8. *Critique of Pure Reason*, B237-8. Kant states explicitly that, 'Thus only can I be justified in asserting, not merely of my apprehension, but of appearance itself, that a succession is to be met with in it'.
9. In Kant's footnote to his new Preface to the B edition of his *Critique*, he states, 'The empirical consciousness of my existence ... is determinable only through relation to something which ... is outside me ... for outer

sense is ... a relation of intuition to something actual outside me ... The reality of outer sense, in its distinctness from imagination, rests simply on its being inseparably bound up with inner experience as the condition of its possibility ... The determination, however, and therefore the inner experience, depends upon something permanent which is not in me, and consequently can be only something outside me, to which I must regard myself as standing in relation. The reality of outer sense is thus necessarily bound up with inner sense, that is I am just as certainly conscious that there are things outside me ... as ... that I myself exist as determined in time ...' BXII. (This translation is that of Eugene T. Gendlin found in Thomas M. Seebohm and Joseph J. Kockelmans, (eds.), *Kant and Phenomenology*, Washington, D. C.: Center for Advanced Research in Phenomenology and University Press of America, 1984, p. 151.) But all these statements are certainly illegitimate on Kant's own standard that one may not possess access to things outside of him, which quite obviously must be things-in-themselves. There is no doubt that his argument is pre-Cartesian since it assumes a certain consciousness of the existence of real (as opposed to imagined) things outside of the subject knower. What is interesting for present purposes is that Kant reveals the grounds for a phenomenological relationship between space and time in this passage although he himself describes the relationship in terms of an empirical order. In this passage, he also indicates his preference for space over time in that his inner, temporal sense is dependent upon that which is permanent outside of him. Kant has neither rendered a pure phenomenological account of space and time nor has he portrayed space and time as equally primordial. He has, however, provided some of the basis for understanding the relationship between space and time without having been fully aware of the implications of his accomplishment. *Cf.*, Robert E. Allinson, *A Metaphysics for the Future*, Part II, Chapter 1, Aldershot: Ashgate, 2001.

10. *Ibid.*, B237. Curiously enough, when one reads Kant's Second Analogy as a definition of pure succession, it would appear to take into account the possibility of time reversal (*Cf.*, Robert E. Allinson, *A Metaphysics for the Future*, Aldershot: Ashgate, 2001, Introduction, n. 5).
11. There is no contradiction between the metaphysical possibility that what happened before at some future time will not have happened (yet). Thus, Kant's defense of causality is preserved in this reading of Kant. For an event to be distinguishable as an event (as over against a steady or enduring state of affairs), it must be capable of being related to a moment of time in which it did not exist. For Kant, the relation was always to a preceding moment in virtue of his preoccupation with the concept of causality which for him implied that the later to be distinguished as later must be related backwards to the earlier. If one wishes simply to define an event as over against an enduring state, it only matters that one can relate the event to a time in which it was not, whether this is past or future. If

time reversal were possible, (with the phenomenologically purified understanding of Kant as rendered in this present, respective work) one could not say that an event once having happened could not not have happened.

12. With respect to cosmological origins, or cosmophysical existence, Hegel does make a connection between space and time when he states that 'The past and the future of time, taken as *existent* in *nature* are space ... ' (emphasis his), *Enzyklopädie*, # 259. This would seem to imply that space is the prior and the subsequent form of time and thus a connection between the two is asserted although the connection is prejudiced toward the primacy of space. The problem is that this remains at the level of assertion rather than definition and is in any case apparently restricted to cosmo-physical existence. Nevertheless, it is of interest that Hegel saw space as the precursor of time and its successor.

Moderato con moto
Easily, but not too slowly

5 The Thalian Fallacy: The Problem of Physics and Philosophy

The majority of physicists in the twentieth century retained a preoccupation with the Thalian search for the ultimate cause or ultimate stuff out of which everything else is made. One candidate for such a cause was the so-called quark family, named after its appearance in James Joyce's *Finnegan's Wake* (hereafter, for simplicity's sake, to be referred to by the single name, quark).[1] With the notion either of an ultimate cause or an ultimate stuff (or unit or packet or energy pattern), the question arises if such an item is transcendent to, or immanent in the universe. If the quark, for example, were transcendent, how could it interact with the perceptual world? If wholly transcendent, how could it interface? If it were transcendent, in what sense could it "cause" the world. By definition, there could be no empirical evidence of any causative interaction since the quark would be outside of the terms of evidence: it would literally be out of this world.

On the other hand, if the quark (or the gluon which is an even more recent candidate than the quark) were already resident in the world, how could it be the cause or the fundamental stuff out of which the world were made? If it were already imminent in the world, it could not produce the world since, *ex hypothesi*, it would already form a part of the world. If it were already part of a formed world system, how could it be the cause of that world system?

If it were to be argued that in some sense the quark were in the world, but only caused part of the world, the problem would still not be solved. As soon as the theoretical entity, or in this case, the quark, were to become perceptible, the question could then be raised, what causes it? If it were a perceptual object, it would be part of the perceptual world and must therefore possess its own origins and conditions for arising in the world. Then what would be responsible for bringing it into being?

The problem appears to be an insoluble one. On the one hand, the quark could not be a part of the world (or the effect) if it were supposed to be the cause of the world. On the other hand, the quark could not be wholly transcendent to the world; or else it could in no sense be a cause of it. Is this "only" a philosophical problem? In other words, is it the case that there must always be a transcendent or a theoretically inexplicable portion in every physical explanation - or, is this a problem of physics? Or, are both physics and philosophy interwoven in this particular case? Perhaps, there must always be a transcendent portion in the explanation to any foundation problem in physics and the use of the term "cause" to describe the interaction must always be in an unknown sense, or in a sense not fully knowable.

In so far as the cause is knowable, it cannot function as the "cause" in the transcendent sense. If unknowable, it equally cannot be said to be a cause (since all known causation is empirical), but it may be an inferred cause in a transcendent sense. In so far as it is knowable as a cause in the empirical sense, it can no longer function as a "cause" in the transcendent sense. In this way philosophy would become an inescapable part of physical explanation. Something unknowable must exist in an ontological sense as a condition for the possibility for something else to be knowable in an empirical sense. In a very loose sense, the problem of ultimate explanation in physics bears an analogy to the Heisenberg Uncertainty Principle, in which one cannot accurately measure both the velocity and the position of a particle at the same time. One of the problems in the case of the quandary of attempting to find the ultimate cause of the physical world is that the world of physics is not a pure empirical or perceptual world, but seems to be a mixture of perception and theory. The world of physics is already a world of refined or inferred perception and hence is a mixed world, which includes theoretical entities both in terms of observations (e.g., traces in Wilson Cloud Chambers) and in the use of language (all language used to describe even purely theoretical entities is borrowed from the perceptual world). The problem is not the problem of attempting to figure out how the transcendent can cause the immanent, or even how the immanent can cause the immanent, but is the problem of attempting to adjudicate between two levels of the perceptual world. The pure world of the physicist is never available either in terms of the content or the language of its description.

Physics and Philosophy

The question entertained above could be reframed in terms of which level (the level of theoretical physics or the level of perception) is primary reality, and which level is in some sense a derivative or dependent reality. If one were to consider the perceived world as the effect of another world, say, the world of quarks, then this world of perception would be a derivative reality. If the world of perception were an effect, it would be in some sense a secondary or a dependent reality.

If the foregoing discussion were to be couched in Kantian terms, the question could assume the following form: how can the relationship between the *ding-an-sich* and appearances be conceived? From a strictly Kantian point of view, one would be forced to take the position that nothing at all can be said concerning this relationship. For Kant, one of the terms in the relation exists on a level outside of appearances, and if one applies the language of appearances to it, one has thereby committed the Kantian fallacy of the transcendental misapplication of the categories. Either there is no relationship at all between the *ding-an-sich* and the appearances, or there is a relationship about which one is forced to be silent. If electrons and electrical phenomena, to choose a classical case as an example of the dilemma, were simply construed to be different modes of description of the same entity, then in what sense would an electron be a more fundamental term of description? Electrons were posited initially as inferred entities, which explained the appearance of phenomena, such as traces or shadows in Wilson Cloud Chambers. Electrons were not conceived of as linguistic descriptions; they were conceived of as real entities. If electrons were simply another way of describing or talking about electrical phenomena, then how would one account for perceiving the "footsteps" left behind in the Cloud Chamber? Something caused the "footsteps" and that something, which caused the "footsteps", was not a phenomenon, that is, not an immediate object of perception. Whether one conceives of the electron as a matter of convenience as a particle or as a probability field around an atom, the problem remains the same. The fact that it is not possible to pin an electron down and that it is a probability wave does not detract from its ontological status. It is, nonetheless, not a purely theoretical construct. A purely theoretical construct cannot leave "footsteps"!

When one plugs a lamp into an electrical source, the lamp has a source of power. The light given off by the lamp is not the same as the source of power with which it is connected and the illumination would not exist in the first place if there were no source of power. As soon as one allows the existence of "theoretical" or "causal" or inferred entities in physics, similar problems that plague Kantian epistemology reappear in the attempts to explain the relationship between the world of physics and the world of perception.

How can two items occupy the same space-time? How can both micro objects and the brown table be in the same space-time? One can attempt to avoid this problem by saying that the brown table is only another level of description of the micro objects, and there is no fundamental difference between the two, in the manner of the solution posed by Searle in *Mind, Brains and Science* (1985).[2] But, the problem with Searle's solution is that the brown table cannot simply be a complicated version of micro objects, because the micro objects are *ex hypothesi* the cause (and not the simplified component part) of the brown table. And, if they are the cause, then either two things are occupying the same space-time - *per impossibile* - or, the very existence of this problem is indicative of a truth about reality, to wit, that reality contains more than one space and more than one time.

In other words, if it were true that electrons, electromagnetic waves (or whatever is the cause of electricity according to the most current theory) were the cause of electrical phenomena, then it would seem to follow, however contradictory or absurd this must seem, that there is more than one space-time.[3] But even this solution, without further emendation, does not altogether solve the problem but only compounds it, since the problem then becomes one of how to causally interact between the different spaces and times, which would appear to require an infinite regress of spaces and times.

For Kant, such questions were to be avoided, since one would be applying a category, such as causality, to the relation between things-in-themselves and appearances and such an application is illegitimate. But, Kant only side-steps the dilemma by saying that it cannot be talked about. This does not remove the problem; it only provides some with a comforting excuse for not needing to solve it.

That the problem remains is also noted in Kant's Transcendental Deductions wherein the Synthesis of Apprehension which must be a mental process since an activity such as a synthesis manifestly takes time, is the constitution (although not the creation) of the phenomenal world, which is the world of time. Kant, in the B Deduction, attempts to avoid any psychologistic language, but this then leaves the entire synthesis on the level of myth: a process that never occurs - which is the tantamount to saying a process exists that is not a process. If the only sense one could make of occurrence is by making reference to time, and this mental "process" cannot *ex hypothesi* take place in time, then it would not be legitimate to make reference to such a process at all. If one were to make reference to a process, then the process must take place, and any activity that occurred would have to occur in time. It could well be argued that this problem is only a pseudo-problem, since it would come into being only if one were to subscribe to Kant's special view of space and time as prescriptive forms. If one were not to subscribe to Kantian views of space and time, the problem would not arise.

However, one must bear in mind that a fundamental analogy exists here with the problem of attempting to reconcile theoretical physics with the world of perception. For Kant, there cannot be two times or two spaces since two times, for Kant, could only exist as times if they were related in terms of time, and then there would only be one time. In other words, for Kant, one could not make sense out of two times, unless the two times were related temporally. But, to relate two times temporally, they must be related inside one and the same time. Otherwise, there would be no sense in which there were two times. If the two times were related temporally, they would *ipso facto* be reduced to one and the same time. Likewise with space, two different spaces (for Kant) could not be different - or exist apart from each other - unless they were separated by one and the same space, and hence there could only be one space.

But, while it may appear that two times could not be different in time unless they were part of the same time, and two spaces could not be spatially distinct unless they were part of one and the same space, it does not follow that the limits of human imagination are rules for the constitution of reality. While according to the present author, it may be the case that a clue to the nature of reality can be found in the necessary relations of human thought, it does not follow from this that there is an exact match between thought and reality. It is only suggested that there can be an analogical resemblance. It is difficult if not impossible to imagine what this resemblance could be. Part of the problem may be due to the fallacy of an observer existing in an absolute space-time standpoint. If one desires to arrive at some knowledge of reality, one must proceed at one's peril.

Is it possible that there could be a second time (or multiple times) which is or are connected to the first time through space, and there could be another space (or spaces) which is or are connected to the first space through time? Would the interdependence of space and time or of multiple spaces and times be of any relevance to the problem of relating the world of physics to the world of perception in the philosophy of physics? Would the existence of a space-time interdependence be of any relevance to the problem of the transcendental synthesis in the Transcendental Deductions and to the difficulties Kant encounters in his First Antinomy in the Transcendental Dialectic?[4]

If there were a problem that does not seem removable on the level of relating the world of physics to the world of perception, then might it be the case that such a problem is reflective of a different configuration of reality? In other words, if a problem were to appear on the level of the difficulty of explaining the relationship between inferred entities in physics and the world of perception, it might reflect that such a problem of "two things existing in the same space" suggests that on a higher level, space and time are connected in ways that are not being currently imagined. It would not be constructive, *à la* Kant, to simply legislate the possibility of approaching a solution to the problem away. That there is a problem must be acknowledged before any constructive solutions to the problem can be conjectured.

For Searle, physics and perception are two differing levels of descriptions of one domain of entities. But, this solution leaves the problem untouched since according to some theoretical physicists, "theoretical" entities in physics cause phenomena.[5] But, they cannot cause phenomena if they are reduced to a different form of description of the phenomena of which they are the putative explanations or causes.

It also cannot be the case, as Collingwood argued, that the world of physics and the world of perception are simply two different language systems such that one speaks in one when one is a physicist, and in another when one is an ordinary perceiver. Such a solution does not explain in what language one speaks when one is attempting to explain the relationship between the existential claims in physics and the world of perception? Does philosophy utilize a different language system from that of physics or ordinary speech? But, the most important difficulty with the multiple language system solution is that the items described in the physicist's language system are supposed to existentially cause the appearance of the items described in the ordinary language system. And, the fact that one may switch between languages of description leaves this fundamental problem untouched.

In the metaphysical exposition of the concept of space-time, ('metaphysical exposition' is employed here in Kant's sense as an exposition which exhibits the concepts as *a priori*), which appears in the previous chapter, it has been argued that space and time are only definable if they are interdefinable. In other words, it is impossible to arrive at a definition of one without recourse to the other. According to the principle that the whole must resemble the part, this conceptual character of space and time may well possess an ontological correlate. Is it possible that if one were to consider the possibility that space-time possesses a dualistic nature that this would shed any light at all on the problem of two items (the entities of physics and those of perception) existing in the same space? What if one were to say that the electron is in the same space as electricity, but from the point of view of the observer, it exists at an earlier time?[6] Is it possible that the postulate of the ontological interconnection of space and time would possess any advantage in addressing the problem of the existence of two different spaces (the space of the world of perception and the space of the world of physics) by reference to time as an intervening medium? Is it possible to say that the micro object causes the macro world by being in the same space, but in a different time?

Perhaps one can reason by analogy from conceptual definitions to onto-physical realities. If space and time were connected conceptually, then by analogy, it could be conjectured that space and time are also connected in reality. The postulation of such a metaphysical realm is not the same as its postulation as part of a speculative metaphysics. The starting point from which such a conjecture has been made is itself phenomenologically verifiable, so that the basis of the inference is not pure conjecture. Metaphysical inferences are justifiable if and only if they are phenomenologically grounded. Then and only then can one make use of the principle of analogical inference that the whole must resemble the part, or that the macrocosm must contain the microcosm.

While metaphysical truths must be epistemologically grounded upon phenomenological grounds, the *existence* of these metaphysical truths is not dependent upon their having been discovered in the course of pure phenomenological investigations. On the contrary, it is the existence of these metaphysical truths that make the phenomenological truth discovery conditions possible in the first place.

What relevance does the special metaphysical postulate of a space-time interconnection possess for other classical philosophical problems? What of the classical question which Kant, in his First Antinomy of Pure Reason, claimed could not be answered, such as, does the world have a beginning in time? For Kant, one could not raise such questions since they were, he thought, unanswerable. For Kant, they were unanswerable because pure reason would entangle itself in insoluble contradictions when it attempted to answer these questions. Does the special metaphysical postulate of the inter-ontological character of space and time shed any new light on the problems that Kant raised in the First Antinomy? Can the special metaphysical postulate of the inter-ontological character of space and time offer any new solution to the question of whether there is a beginning to the universe or might it be the case that it discloses rather a different explanation for the incapacity of reason to solve such a problem?

Could it be argued that while it seems self-contradictory to understand how something can begin in time, that there is a gain in comprehensibility if one attempts to conceive of time's origin in space? While it seems self-contradictory to speak of an end of space in space, could it be said that space possesses or will possess an end in time? Is it necessary to say, with Kant, that reality can have no description? Is it possible to say that reality is describable analogically in such a way that a necessary truth on a conceptual level is likely to have a parallel ontological correlate on the other? One can refer to the inferred parallel ontological correlate as a special metaphysical postulate. Space can come to birth in time and time can come to birth in space. Is this a possible new metaphysical solution to Kant's question?

It may well be argued that what is suggested here is no solution at all, since earlier it was argued that space and time were inter-definitional. If space and time were inter-ontological, then how could it be that one could be born of the other? And, if one were to exist first, then *a fortiori* it would exist without the other. It is difficult indeed to answer this conundrum. Could it be that this conundrum arises because of the presupposition of a single space-time continuum? If reality were constituted by a plurality of space-time continua, then the space of one space-time continuum would not need to co-exist with the time of the same space-time continuum. The space of one space-time continuum would co-exist with the time of a different space-time continuum. Nevertheless, it could be argued that this would entail an infinite number of space-time continua and in the end would only result in the same problem arising all over again.

One cannot represent space without time or time without space whether or not one may make reference to the time or space of a different space-time continuum. Each time that one attempts to represent space, one finds oneself ineluctably drawn into the inclusion of time. Each time that one attempts to represent time, one finds oneself ineluctably drawn into the inclusion of space. In the order of understanding or explanation, one always understands space and time together. Whensoever one attempts to understand the nature of space, one must always make reference to the nature of time. Whensoever one attempts to understand the nature of time, one must always refer to the nature of space. This conceptual interdependence cannot be avoided. In terms of relativity theory, the question whether an independent velocity or position can exist apart from the question of the subject knower is a question that makes no sense. Perhaps, it is the case that the question of the origin of space-time may be one which would always occur to a subject questioner and may be a pseudo-issue outside of the frame of reference of a subject questioner. The answer, so to speak, is in the question. Does this non-answer as an answer shed any different light on the question than the non-answer as an answer proposed by Kant?

At this point, it may well be said that reality as it does exist exceeds conceptual comprehensibility. While this is an omnipossible hypothesis, it is important that the philosopher does not seize upon such a possibility before its time. Time and space are intertwined in a fundamental twoness that appears as an infinite process. But, if infinite, do not all the same questions reappear under different descriptions? Perhaps so, and this may be reflective of the need to include this problem as one of the 'Eternal Problems of Philosophy'. But, the problem is not quite the same one as it previously appeared to be. It has been raised to a higher level and dialectically purified, refined and amplified so that it contains different elements and is in reality a different problem.

If space and time were intertwined, it may be that each attempt at understanding one member of the pair inevitably involves the subject knower in the understanding of the paired other. In this sense, it is impossible to understand the beginning or ending, since each member of the pair would appear to precede the other unless they were to come into being together. But, the coming into being is itself the question, not the answer.

The only proper sense one may be able to make out of this, as a philosopher, is that in the attempt to comprehend the limits of space or time, one must perforce employ the concepts of space and time. So far, this solution does not seem to differ from that of Kant's except it includes space and time together instead of treating them separately. However, it is also possible to suggest that the reasons why space and time must be employed are different from those reasons proposed by Kant. In the explanation offered here, it is because of the nature of the fundamental inter-ontological existence of space and time that the subject knower, as a creature of the universe, cannot avoid utilizing space and time in her or his explanations. Even in the attempt to explain the beginning or ending of space-time, the subject knower cannot help but think space and time together and hence is ineluctably involved in a self-contradiction. It may be that it is owing to the nature of the structure of the universe and not to the incapacity of reason that the subject knower cannot arrive at a completely satisfactory solution to the problem. One could say that there is a dual reason for the inability of the subject knower to resolve this question. One reason is that space and time cannot be conceptualized unless they are conceptualized together. As a result, it is impossible to argue that space is the origin of time or that time is the end of space. It may be argued further that this inability to conceptualize space or time without the other is due to the nature of their fundamental interconnection in the universe and not due to the psychological incapacity of the subject knower or to the limits of pure reason.

Notes

1. The name 'quark' was given to elementary particles (which superseded protons and neutrons as 'elementary' particles) by the Caltech physicist Murray Gell-Mann, who was awarded the Nobel Prize in 1969 for his work on them. The origin of the name is, according to Hawking, James Joyce's enigmatic sentence: 'Three quarks for Muster Mark!' According to Gell-Mann, he first thought of the sound and later found the word in Joyce. While the word 'quark' was supposed to be pronounced like the English word 'quart' except with a 'k' at the end instead of a 't', it is more commonly pronounced to rhyme with the English word 'lark'. *Cf.*, Stephen Hawking, *A Brief History of Time*, New York: Bantam Books, 1989, p. 69. At first there were three quarks and then four. Poetic names were chosen for the quarks such as 'strange' and 'charm'. These names were referred to as flavors. The quarks were postulated to appear in different colors. The quark family grew to eighteen. The quark model as described here (considerably oversimplified) underwent more and more modifications but still presented difficulties. One difficulty was that the quarks, though postulated as physical existences, were never observed. In his afterword to the second edition of the *Tao of Physics*, 'The New Physics Revisited', Fritjof Capra suggests that a bootstrap theory of particles has been devised that can account for the quark structures without needing to postulate the physical existence of quarks. In this explanation, quark structure 'appears as a manifestation of order'. *Cf.*, Fritjof Capra, *The Tao of Physics*, Boston: Shambhala, 1991, p. 318. However, it is not clear how the concept of order does away with the need for physical existence. The word 'manifestation' in Capra's phrase surely carries the meaning of to manifest or to show itself. If a quark is manifest, then it is visible in some sense. Indeed, what sense can be made of a structure which contains no components? In any case, order is an observed or descriptive phenomenon; it is not a causal explanation. If order were a causal explanation, and the order described were the order of the whole, the part would thereby be determined by the whole, and it is difficult to see how determinism could be avoided. Plainly,

it cannot be the manifestation of the part for that which possesses no physical existence cannot be the cause of itself.

2. According to Searle, 'In the case of liquidity, solidity and transparency, we have no difficulty at all in supposing that the surface features are *caused* by the behaviour of elements at the micro-level, and at the same time we accept that the surface phenomena *just are* features of the very systems in question' (emphasis his). *Cf.*, John R. Searle, *Minds, Brains and Science*, Cambridge: Harvard University Press, 1985, p. 21. However, it is the point of this chapter to show that there is indeed a difficulty in accepting the notion that that which is a cause is at the same time an effect.
3. This is a simplified and condensed form of explanation. Electricity has two forms, current and static electricity. Current is one manifestation of the flow of electrons. The cause of light is not, strictly speaking, the flow itself but rather the jumping of one electron from a higher to a lower level of energy level in an atom, thereby losing energy which is regained by the emission of a photon, which is the phenomenon of light. The point, however, remains the same. There is still a difference between the electron (or its energy loss) and the effect, which is light. It is the energy loss that is the cause of light in that light cannot exist unless there is this physical change. The energy loss must take place prior to the phenomenon of light. The phenomenon of light cannot be reduced to a description of the emission of a photon because the emission of a photon is itself a result of the previous jumping of an electron. A causal process has taken place that results in the production of light. The process cannot be reduced to its elements or constituent parts. The active nature of the process and its temporal order are keys to its existence and functionality. The author is indebted in this connection to discussions with the physicist, Finn Aaserud, Director of the Niels Bohr Archives in Copenhagen. It may be said quantum jumps are not necessary to radiate light and that an electron will radiate if it is accelerated fast enough. It still remains that it would seem to be the case that a causal process must exist whether or not one possessed recourse to the explanation of quantum leaps. Even it is argued that any accelerated charge will generate light, nonetheless, the acceleration must precede the light effect. There is still a causal order and a causal activity. The electron must do something before there is the phenomenon of light. The action of the electron must precede the light effect. The light effect is the result of something that the electron has done and would not exist apart from that fact. The question of quantum leaps, nevertheless, is an exciting point of philosophical conjecture. It is important to note that the very notion of a quantum jump is not in itself a clearly defined notion. In a famous conversation between Schrödinger and Bohr, Schrödinger asks Bohr whether the jump is gradual or sudden. He is led by his own puzzlement over the inexplicability of the notion of quantum jumps to exclaim, '... the whole idea of quantum jumps is pure nonsense'. *Cf.*, Werner Heisenberg, (trans. Arnold J. Pomerans), *Physics and Beyond, Encounters and Conversations*, New York, Evanston, San Francisco,

London: Harper & Row, 1972, p. 74. Bohr replies that Schrödinger's puzzlement over the notion of quantum jumps, '... does not prove that there are no quantum jumps. It only proves that we cannot imagine them, that the representational concepts with which we describe events in daily life and in classical physics are inadequate when it comes to describing quantum jumps. Nor should we be surprised to find it so, seeing that the processes involved are not the objects of direct experience'. To this, Schrödinger makes the interesting reply that, 'I don't wish to enter into long arguments about the formation of concepts; I prefer to leave that to the philosophers'. From this tiny excerpt of a brilliant conversation one can infer at least four major conclusions. First of all, the very notion of a quantum jump, while posited as a key solution to a problem in physics is itself a metaphor and describes an invisible process. Secondly, the metaphor of a jump is itself not problem free. Thirdly, that one can attempt to resolve problems in physics or absolve oneself from the need to resolve them by appealing to the standpoint that all representational concepts or linguistic descriptions are bound to be inadequate and hence faulty. (This 'solution' or 'absolution' is itself inadequate since some representational concepts or linguistic descriptions are better than others.) Fourthly, that when one is attempting to describe the most fundamental foundations of physics, one is inevitably involved in philosophy. If at this point the physicist (as Schrödinger) appeals to philosophers, then it is important that philosophers become partners to physicists in the attempt to sort out the most basic dilemmas which emerge from the attempt to explain the basic operations of the universe.

4. It is of interest to point out that Einstein himself saw some kind of undissolvable connection between space and time. According to Andrew Paul Ushenko, writing in his article 'Einstein's Influence on Contemporary Philosophy', 'Einstein endorsed Minkowski's dictum that "henceforth space in itself and time in itself dissolve into shadows and only a kind of union of the two retains an individuality" '. *Cf.*, Paul Arthur Schilpp, (ed.), *Albert Einstein, Philosopher-Scientist*, The Library of Living Philosophers, Vol. VII, La Salle, Illinois: 1991, Fifth Printing, p. 637. (Sir Roger Penrose's rendition of Minkowski's famous dictum announced in his 1908 lecture at the University of Göttingen is: 'Henceforth space by itself, and time by itself, are doomed to fade away into mere shadows, and only a kind of union of the two will preserve an independent reality.' *Cf.*, Roger Penrose, *The Emperor's New Mind*, New York: Oxford University Press, 1989, 1990, p. 193.) It seems that philosophy has taken some time to catch up with physics in that it is only just now that a conceptual proof of the interdependence of space and time has been offered. It is important to remember that the philosophical proof is not dependent upon the scientific idea or vice-versa. What is of interest to note is that if there is a unity of nature, and man is part of that nature, then what man can discover by the analysis of nature should parallel what man can discover by the analysis of concepts. If there were a contradiction between the findings of the realms

of philosophy and science, there would be cause for worry. That there is a convergence of truths obtained from different lines of discovery is a cause for celebration.

5. Analogously, Penrose argues in his debates with Hawking that one cannot be satisfied by a purely epistemological explanation in physics, but one must hold some realist position. *Cf.*, 'The Nature of Space and Time', *Scientific American*, July 1996, p. 49.
6. This is not the same problem as that of Schrödinger's cat. The problem of Schrödinger's cat is a pseudo-problem that has arisen because of the construction of a thought-experiment without giving sufficient attention to distinctions between ontological and epistemological considerations. Here, philosophical distinctions may have aided physics had they been considered before, rather than after empirical experiment (to be fair, it is a thought experiment rather than a scientific experiment proper that is the basis of the so-called Schrödinger's cat dilemma). A cat is gruesomely placed in a box. If detected by photon emission, such emission triggers a gun, which then kills the cat. If not detected, then the cat is alive and well. Thus, one is presented with the putative paradox of a cat that is both alive and not alive at the same time. In this case, the dilemma is a false one, which is the result of the experiment being falsely conceived. One is conflating the level of the cat as perceived of as existing in itself (purely ontologically) with the level of the cat as perceived after experimentation (which includes an epistemological element) - which also changes the original experimental datum, which is the cat as perceived of apart from being perceived of by photon emissions which trigger a gun. Hence, the alleged dilemma, which ensues, is the result of mixing ontological and epistemological levels of existence. The cat as it exists in itself does not belong to the same category of existence as the cat as perceived by photoemission linked to a gun. There is no real existence of a cat which at the same time is alive and not alive. There is only the conjectured existence of a cat as it exists apart from being known and a conjectured existence of a cat as it exists as known to exist via photoemission. These two levels are not comparable and hence cannot be fruitfully compared. Thus, Schrödinger's cat dilemma is no dilemma at all, if the present author may be permitted to let the cat out of the box! One may find a reasonably clear depiction of the supposed paradox (not recognized to be a pseudo-paradox) in Stephen W. Hawking and Roger Penrose, 'The Nature of Space and Time', *Scientific American*, July 1996, p. 47. John Gribben argues that Schrödinger thought up the experiment to find a chink in the armor of the strict Copenhagen interpretation of reality. *Cf.*, John Gribben, *In Search of Schröedinger's [sic] Cat, Quantum Physics and Reality*, London: Corgi Books, 1984, p. 205. However, if the argument presented above is correct, then the cat dilemma sheds no light at all on the Copenhagen/Einsteinian question.

Allegretto piacevole
Not too fast, with graceful and equalized movement

6 The Complementarity Principle, The Sixth

By nature, philosophers seem to be divided up into two instinctual kinds: one kind tries to figure everything out; the other kind tries to explain why it is not possible to figure everything out. In the approach taken here, there is some accommodation to both instincts. While there is some attempt made to set limits for philosophy, so that speculation will not be totally rampant and unverifiable, on the other hand, within those limits, there is also some attempt made to show how a certain level of knowledge is possible which is filled with meaning and content.

The basic postulate is that there is a complementary relationship between what can be known to be true in terms of the discoveries made within phenomenological epistemology and the nature of the universe itself. The reason that such a complementary relationship is posited to exist is itself based on the principle that the macrocosm, having generated itself from the microcosm, must contain the microcosm, in some form, within itself *qua* macrocosm.[1] This is a way of putting the whole-part argument of metaphysicians, which might sound abstract to some since it is based on the logic of concepts, in the language of evolution. If this principle is applied to what is being carried out here in terms of the relationship between phenomenology and metaphysics, then what is being said is that there is some kind of isomorphism that exists between the relationship between what can and cannot be known in pure phenomenological knowledge, and what can and cannot be known regarding the structure of the universe.

There is an analogical resemblance between the kind of work carried out here and the kind of work conducted by philosophers in the past. Kant's thesis in his First Antinomy was that the world had a beginning in time and was also limited in space. His antithesis was that the world had no beginning in time and was not limited in space. He managed, so he thought, to prove both thesis and antithesis and thus displayed reason as being in conflict with itself. Reason was powerless to solve such a dilemma since it had attempted to address the question of the nature of the thing in itself.

What sense can one make of Kant's arguments when it is apparently the case that according to co-temporary developments in physics, proofs have been developed that demonstrate that both time and space have a beginning?[2] Whether or not it makes sense philosophically to say that time has a beginning, is it permissible to argue with scientific proofs? According to the British physicist, Hawking, not only is it true that time and space have a beginning, but that it is necessary that they do.[3] Shall it then be said that the conception that a beginning requires a pre-existent time is simply a conceptual obstacle and has no bearing on the workings of physical reality? In other words, it is irrelevant whether or not a beginning to time is comprehensible to the human intellect. The world does not have to package itself in a conceptually comprehensible form. What right does the subject knower possess to assume that the universe must function in strict correspondence with the limits of her or his capacities to conceive?

This resolution of the Kantian-modern physicist dilemma has a certain palatability to it. Whatever the philosopher might consider about the conceptual impossibility of a time that begins outside of time, in terms of physical fact, the time that can be measured can only be said to begin (simultaneously?) with the beginnings of space (shades of Hegel). Once again, the indissoluble wedding of space and time is rediscovered. If the celebrated Aristotelian definition of time is given a slight spin, time becomes the number of spatial motion. To be sure, this aspect of time refers only to its mathematical and not its experiential side, but in any event, it does show the actuality of the anticipated isomorphism between physical space and time and phenomenological space and time. Time only exists in the expansion of space, and space itself can exist only in time. There appears to be a wondrous parallel between the space and the time of the physicist and the space and time of the phenomenologist.

The Sixth

What does the foregoing have to do with figuring everything out or explaining why everything cannot be figured out? The philosopher, left to her or his own devices, cannot figure everything out. She or he must utilize evidence discovered by the physicist in order to make her or his theory more complete. On the other hand, she or he can then take this evidence from the physicist and correlate it with the philosophical evidence (e.g. the introspective evidence concerning the relationship between space and time) in order to speculate properly about matters that go beyond the limits of the competence both of the physicist and the phenomenologist proper. This is the new arena for metaphysical exploration: the realm of analogical resemblance, or, the Sixth.

What can be said, then, about the Sixth that is a result of a comparison between the domains of philosophy and theoretical physics? Perhaps, it can be said that the universe always displays an equiprimitive dipolar characteristic. If it is the case that space and time are not reducible to each other, but can only be understood by reference to each other in terms of phenomenological space-time, and this relationship holds true in the physical universe as well, what appears to have been discovered is a metaphysical principle of the irreducibility of a dipolar phenomenon. What implications this has for physics, in turn, is difficult to foresee at the moment. But, it appears to be a speculation that is now at least doubly grounded, both on the phenomenological and the physical level. One possible implication would be that if a unified field theory were possible, then any single explanation which subsumed all natural phenomena under it would necessarily be composed of at least two irreducible elements. If dipolarity is primitive, then any ultimate physical explanation must be composed of at least two fundamental elements or constituent parts which are not ontologically independent or conceptually explicable in isolation from each other, but require each other for both their ontological existence and their intelligibility. In this sense, Bohr's complementarity principle as an explanatory principle may possess implications which go beyond the notion that the wave and the particle theory of matter are both necessary and irreducible explanatory theories.[4]

It may be argued that this is nothing more than serving up old wine in new bottles. The dipolar nature of reality can be found in the *I Ching* or the *Book of Changes* as evidenced by the discussions of *Yin-Yang* polarity.[5] While what is advanced here is not totally new, the difference is that what is presented here is not the product of creative speculation, but has been first grounded in phenomenology and physics. Instead of the pure imposition of a theoretical structure on the world, the fundamental, irreducible dipolar character of the world is found originally in phenomenological space-time and then and only then also posited as possessing a possible application to physical space-time.

In other words, more evidence can now be given to show that the particular mythical framework (the dipolar nature of reality) presented in the *I Ching* may be characteristic of the universe. The methodology of demonstration employed here is not to be identified in any respect with the methodology of beginning with a chosen framework and then attempting to find it exemplified everywhere in the mode of the rationalist metaphysician or in the popularizing counterparts of the rationalist metaphysician to be found in the guises of various books which seek to demonstrate the relationship between physics and Eastern philosophy. In contrast, the method followed in this work has been to begin by marking out some limits for knowledge, and then seeing if one can transgress those limits by utilizing what can be known within those limits, and comparing that with what can be known within the limits of physical knowledge. Thus, the *path* to what is discovered is quite different from simply starting out with an abstract principle. Even to start out with "Eastern" philosophy or the relationship that exists between Eastern philosophy and co-temporary physics is to start out with an abstract principle. The problem with either (or both) of these standpoints is that they may be mistaken. If one arrives at the correct results by accident, the method employed to find those results cannot be reduplicated to discover other results since the relationship between means and end is accidental. If, however, there is a definite connection between method employed and results obtained, then one may place a greater reliance in the future on the method employed in the search for other results.

All that can be said on behalf of the present method is that it is a proposal for a new *point d'appui* for philosophical work. What has been attempted here is only the very first step. But a beginning is a beginning, and one must start somewhere. From the double grounding of the inextricability of space and time one can make a general inference to the dipolar nature of the universe. It does not follow from this that all pairs of opposites exist or that everything that exists must exist in tandem. It does not follow from this that all pairs bear symmetrical relations to each other. It only follows that if one looks for complementary pairs, this may be of particular assistance both in terms of research in physics and in terms of introspection in phenomenology. The results of the present speculation thus can serve as regulative principles of research for researchers within each discipline. But, it must not be assumed that a dipolarity exists between all seeming pairs or even that a pair that seems to exist as a pair is really a pair. For example, in the ethical realm, it may not be correct that a dipolar distinction exists between good and evil. This may in fact be a pseudo-pair. But this does not detract from the notion that ethical research may be enhanced if one looks for a real pair, such as the relationship between positive and privative that exists between love and fear. Pointers to future research in the ethical realm can be found in the sequel.

One final example. Niels Bohr originally arrived at his conception of complementarity because of his puzzlement over the dilemma in quantum physics that micro objects sometimes seemed to behave like waves and at other times seemed to behave like particles.[6] As a resolution, he proposed that both principles of explanation were required in order to arrive at a complete theory of matter. One applied one principle under certain circumstances and the other under other circumstances. He did not wish to allege that matter itself was ultimately dual in nature but wished to advance his principle of complementarity as a principle of explanation, not as one of physical description.[7] If, however, one thought, unlike Bohr, that for a coherent explanation, one would need to consider that atomic matter could manifest itself in two forms, the Thalian conviction that the ultimate stuff of the world can be reduced to one thing would be challenged.

It seems undeniable that atomic matter behaves under certain conditions as one structure and under other conditions as another. Matter, in this sense, is not ultimately reducible to one form or arrangement. Whether the wave or the particle theory of matter finally are replaced in physics by more sophisticated models is not the point at issue here. From the standpoint of dipolarity, what appears to be the case is that there will always be at least two modes of appearance, and what is more, that these two modes are equiprimitive or equiprimordial. It may not be the case that the wave or the particle mode are the final models accepted by physics, but if what is said here is correct, then it is very likely that physics will always find at least two complementary explanations necessary to describe the structure of the universe. It may even be the case that the wave and particle behaviors of matter are related in some fundamental way to each other such that under certain conditions matter behaves as one and under other conditions as the other.

Rather than finishing this discursus into metaphysics with a regulative note to the further research of physicists, one aspect of the complementarity principle that has the most important regulative note attached to it for the further research of philosophers may be emphasized. The fundamental twoness of reality is not a twoness that is in conflict with itself; it is a twoness that completes itself. Here, the fundamental Western cultural bias that matter must ultimately be reducible to conflictual, individual things may be replaced by an Eastern cultural bias that matter must ultimately be reducible to inherently complementary existing pairs. What is of special interest is that in this case the world of Western physics lends its special support to the Eastern stress on the ultimate harmony of human relations. While earlier lessons were being drawn from philosophy for physics, here lessons from physics may be taken for philosophy.[8] The wave-particle theory of matter is not, of course, the ultimate concern of philosophy. But, the fact that matter must always be conceived of under two theories which are not in conflict with each other speaks to the issue that at the heart of the universe, there is harmony, not conflict. This is not the same as a Heracleitian underlying harmony that requires or justifies surface strife. What physics appears to be showing is that two theories require each other and neither theory is in conflict with the other. There is a plurality which is ultimate, but the relationship between the individual members in that plurality is one of cooperation, not strife. That this is a lesson that comes from the core of physics is highly instructive. That it is a lesson that is one of the first lessons which is also derived from the phenomenological exploration of the nature of space and time is additionally instructive. Throughout reality and throughout introspective reflections on reality, and further reflections on one's reflections on reality, one continues to discover the operation of the complementarity principle at work.

Notes

1. There is already evidence that the microcosm contains the macrocosm in some form. Every cell of an organism contains in its DNA the blueprint of the whole and can reproduce it precisely. Every packet of radiation in a hologram contains a picture of the whole. In fractual mathematics, much of nature consists of structures whose patterns recur at every level no matter how small or large. *Cf.*, Margaret J. Wheatley, *Leadership and the New Science: Learning About Organizations from an Orderly Universe*, San Francisco: Berrett-Koehler Publishers, 1992.
2. *Cf.*, John Boslough, *Beyond the Black Hole, Stephen Hawking's Universe*, UK: Collins, 1989.
3. *Ibid.*, p. 46. One wonders if even within physics one should totally countenance the notion of a beginning to time since this would seem to imply an absolute time or the simultaneous occurrence of space (the starting point of the universe) with time. The Einsteinian question would seem to be, for whom? However, according to Stephen Hawking, both he and Roger Penrose showed that Einstein's general theory of relativity implied that the universe must have a beginning and, possibly, an end. *Cf.*, Stephen Hawking, *A Brief History of Time*, New York: Bantam Books, 1989, p. 36. Hawking, it seems, is not always consistent on this point. For example, he states that, 'We have known for twenty-five years that

Einstein's general theory of relativity predicts that time must have had a beginning in a singularity fifteen billion years ago'. One paragraph later he states that, 'Jim Hartle and I suggested that the universe may not have any beginning or end'. *Cf.*, Stephen Hawking, *Black Holes and Baby Universes and Other Essays*, New York: Bantam Books, 1994, p. 46. He attempts to extricate himself from this seeming dilemma by differentiating between real time in which the universe has a beginning and imaginary time in which the universe has no beginning and no end. *Ibid.*, p. 172. But does imaginary time count as temporality? To further muddy the waters, recent measurements of super novae indicate that perhaps the universe is accelerating and that therefore there would not be an end to it. However, this argument is not strong since the current period of acceleration could stop and be replaced by a period of deceleration.

4. This is the argument of the complementarity principle in atomic physics as advanced by Niels Bohr, which he introduced for the first time in public in 1927 in Como, Italy in a lecture on 'The Quantum Postulate and the Recent Development in Atomic Theory' at the International Physical Congress. *Cf., Niels Bohr, Essays 1932-1957 on Atomic Physics and Human Knowledge, The Philosophical Writings of Niels Bohr*, Vol. II, Woodbridge, Connecticut: Ox Bow Press, 1987. According to Heisenberg, Bohr developed his ideas on complementarity while on vacation in Norway. *Cf.*, Max Jammer, *The Conceptual Development of Quantum Mechanics*, New York: McGraw-Hill Book Company, 1966, pp. 347, 351. (For Pais, it was the choice to use the word 'complementarity' that occurred during one of the sailing trips. *Cf.*, Abraham Pais, *Niels Bohr's Times, In Physics, Philosophy, and Polity*, Oxford: Clarendon Press, 1991, p. 311). Simply put, the complementarity principle states that an exhaustive description of quantum phenomena is possible only through recourse to two, mutually exclusive sets of classical concepts. One cannot employ either the expressions 'Either-Or' or 'Both-And' to Bohr's complementarity principle since the former implies that one of the explanations is enough and the latter implies that both can apply simultaneously. Finn Aaserud, Director of the Niels Bohr Archives greeted the author's description of 'Either-And' with a positive response, when it was put forth in a private conversation on 20 October 1995 at the Archives. Heisenberg's uncertainty principle is a special case of the complementarity principle. *Cf.*, also, Sandro Petruccioli, *Atoms, Metaphors and Paradoxes, Niels Bohr and the construction of a new physics*, Cambridge: Cambridge University Press, 1993.
5. *Cf.*, Chung-ying Cheng, 'Chinese Metaphysics' in Robert E. Allinson, (ed.), *Understanding the Chinese Mind: The Philosophical Roots*, New York: Oxford University Press, 2000, Tenth Impression. While Bohr considered the complementarity principle to be a strictly epistemological one, and not ontological, when knighted, on Hanna Kobylinski's (the wife of Bohr's right-hand man, the physicist Rosental) advice, he chose the *Tai-ch' i* (The Great Ultimate) symbol for the emblem of his coat of arms with

the accompanying motto, *Contraia sunt complementa* (Opposites are complementary). In addition, his favorite quotation from poetry was from Schiller's *Sayings of Confucius*: 'Only wholeness leads to clarity'. *Cf.*, Henry J. Folse, *The Philosophy of Niels Bohr, The Framework of Complementarity*, Amsterdam: Elsevier Science Publishers, 1985, p. 54. Furthermore, the dragon, symbolizing wisdom was originally proposed as the symbol for Bohr's coat of arms when Bohr received the Order of the Elephant (he was the first non-Royal to receive this honour). But this was turned down on the grounds that the dragon was not a heraldic animal. As a counter-offer, it was proposed that stars be placed inside the four corners of the figure on the shield in the center of which the *Tai-ch' i* symbol was to be placed, but Bohr turned down this idea on the grounds that stars could not be inside the cosmos. Thus, unlike all the other coats of arms in Fredericksberg Castle, the emblem on Bohr's shield has nothing in any of the corners of the figure in which the *Tai-ch' i* symbol is placed.

6. There is considerable discussion as to whether Eastern philosophy influenced Bohr's discovery of complementarity in physics. In a private discussion with his son Hans at the Niels Bohr Institute in Copenhagen on 17 October 1995, the author was told that Bohr had read Lao Tzu in a Danish translation. The idea of the complementarity of opposites is advanced in the *Tao De Ching*. One can trace the influence of Western philosophy on Bohr's discoveries in quantum physics as well if one takes into consideration that Bohr attended lectures and read works of Harald Höffding who expounded Kierkegaard's thought including the concept of the leap. According to Höffding, 'Something decisive occurs always only by a jerk, by a sudden turn which neither can be predicted from its antecedents nor is determined by these ... It seems to be clear that if the leap occurs between two states or two moments, no eye can possibly observe it, and since it therefore can never be a phenomenon, its description ceases to be a description'. Furthermore, 'causality cannot be described'. *Cf.*, H. Höffding, *The Problems of Philosophy*, G. M. Fisher (trans.), New York: Macmillan, 1905, pp. 76, 83-4. (Kierkegaard himself in turn traces his discovery of the leap to a reading of Leibniz' *Theodicy* in Gottshed's 1763 edition. In a journal entry he notes, 'Leibniz makes the very important observation S 212 that there are great difficulties involved in the inference of quality from quantity ... '.) Jammer points to a passage in William James which explicitly discusses an example and an explanation of complementarity in an experiment performed by Janet, and Bohr himself was very struck by the philosophy of James. Höffding had visited James in America in 1904 during which time Bohr was Höffding's immediate student. Bohr was influenced by discoveries in mathematics as well. In 1910 Bohr was already influenced by Riemann's notion that cognitions must be arranged into planes that cannot be compared. *Cf.*, Max Jammer, *The Conceptual Development of Quantum Mechanics*, New York: McGraw-Hill Book Company, 1966, pp. 172-180, 345-359. Folse traces philosophical influences even further back to Bohr's student days pointing

to the evidence that friends from those days said when he unveiled his new viewpoint twenty years after his student days that 'this was nothing but what he had been saying all along'. *Cf.*, Henry Folse, *The Philosophy of Niels Bohr, The Framework of Complementarity*, Amsterdam: Elsevier Science Publishing Company, 1985, p. 43-5, 50-53. Indeed, even as a young child Bohr listened to discussions between his father and Höffding held regularly in the Bohr home. Folse thinks, however, that there is no evidence for a connection between Kierkegaard's leap and Bohr's notion of quantum jumps, but Jammer does provide such evidence. *Cf.*, *Ibid.*, p. 47. In any case it would appear to be fair to say that engagement with philosophical ideas in general and complementarity in particular predisposed Bohr to look for, notice and emphasize complementarity as a solution in response to certain found dilemmas in atomic physics. This would appear to amount to a strong testimonial to philosophical ideas as regulative principles for scientific research. In addition, new evidence has come to light from a letter written by Bohr himself, discovered in 1998, which is a reply to a letter of inquiry from a Svend Hugo Jürgensen, a teacher from the Danish town of Aalborg. The teacher had sent Bohr a manuscript entitled '*Tao De Ching* and the Idea of Complementarity'. Bohr's reply, dated 26 March 1958 begins, 'I thank you for your letter and the enclosed little note about the *Tao De Ching*, which I have read with great interest. I believe what you say about the old Chinese philosophy is in many ways quite to the point. In my youth I received a beautiful impression of it through Ernst Møller's book "*Oldmester*", and at a visit to China twenty years ago I learned how highly the memory of Lao-Tzu is still valued'. Since Møller's book was first published in 1909 (the year Bohr turned 24) and Bohr by his own account was in his youth, it is clear that his knowledge of Chinese philosophy preceded his discovery of the complementarity principle in physics (1927). While this does not prove that he learned his idea of complementarity from Chinese philosophy, it is nonetheless a document of great interest. One must bear in mind that '*Oldmester*' is a Danish rendition of the name Lao-Tzu and the book itself is a translation of and a commentary upon the manuscript left by Lao-Tzu. This may well have been the manuscript Bohr's son, Hans, was referring to when, in 1995, he told the present author that his father had been acquainted with Lao-Tzu in a Danish translation. This would then confirm the anecdotal evidence the present author received from Bohr's son. Bohr's acquaintance with Lao-Tzu was both early and authentic. The present author is indebted to the physicist, Finn Aaserud, Director of the Niels Bohr Archives, both for bringing this letter to his attention and for providing the translation above. The original letter can be found in the 'Niels Bohr General Correspondence'. Pais, however, strongly objects to the concept that either Chinese philosophy or Western philosophy influenced Niels Bohr. *Cf.*, Abraham Pais, *Niels Bohr's Times, In Physics, Philosophy and Polity*, Oxford: Clarendon Press, 1991, p. 24.

7. Heisenberg quotes Bohr to have said in 1952 that '... all those who have truly understood quantum theory would never dream of calling it dualistic'. *Cf.*, Werner Heisenberg, *Physics and Beyond, Encounters and Conversations*, London: George Allen & Unwin, Ltd., 1971.
8. It is of special interest to note that Bohr himself held that the greatest significance of the ideas of physics for philosophy lay precisely in the implications that new ideas of physics possessed in challenging the foundations of our most fundamental concepts. Rasmussen points out that, 'In the very first sentence of his 1958 contribution to Klibansky's, *Philosophy in the Mid-Century* Bohr said, "The significance of physical science for philosophy does not merely lie in the steady increase of our experience of inanimate matter, but above all in the opportunity of testing the foundation and scope of some of our most elementary concepts" '. *Cf.*, Erik Rasmussen, *An Essay on Fundamentals of Political Science Theory and Research Strategy*, Odense University Press, 1987, p. 124.

Andante
Quiet and serious

7 The Analogies of Experience

The Grounding of Metaphysics and the Ground of Metaphysics

As Kant discovered some time ago, a yearning for metaphysical knowledge cannot be expunged. No matter how powerful the stricture is against metaphysics, there is a built-in urge to possess such knowledge, however dim and however loosely verifiable such knowledge might be in the end. A new task, which is here being set, is to first delimit the field to avoid rampant and irresponsible speculation, and then to engage peripherally in speculative metaphysics in a limited way, so long as some responsible checkpoints for an extension of knowledge to the whole can also be provided along the way.

The problem with metaphysics in the classical sense (whether descriptive or revisionary in Strawson's sense) is that it lacks, on the whole, reasonable checkpoints along the way. However much the systems of Hegel or Whitehead are to be admired and succeed in arousing one's philosophical wonder, there remains an air of unreality about them. One cannot know if they are really true. In the end, they seem more like poetic statements than philosophical ones, and thus offer resistance to genuine and full-hearted commitments.

F. H. Bradley's nefarious comment on metaphysics suggests a point of departure: 'Metaphysics is the finding of bad reasons for what we believe on instinct, but to find these reasons, is no less an instinct.' In reconstituting Bradley's notorious statement, the task here is to find good reasons to account for the possibility of what appears both as an inherent disposition in human nature, and also appears to extend in its reference to the world, or the universe as a whole. The reconstitution of Bradley's remark would read: metaphysics consists in finding *good* reasons for what in phenomenological experience is discovered to be true.

If the starting point for metaphysics were to be a set of 'good reasons', one would return to the camp of the rationalist metaphysicians in attempting to fit both human nature and the world into a systematic structure. This is not the starting point adopted here. On the other hand, one cannot begin with instincts *per se*, since one could be mistaken about what instincts are possessed. In order to better sort out what are true instincts and what are not, one may be referred to Part II, Chapter One, below. A beginning with instincts may provide a clue to the metaphysical foundations of ethical knowledge to be discussed in the sequel.

There are two analogical levels. The first is that one can, while still within the phenomenological standpoint, discover the limits of *a priori* knowledge. However, just outside of phenomenological knowledge proper, one can find some suggestive corresponding knowledge in the physical world. For example, it so happens that while mathematical knowledge is that which takes its rise from *a priori* intuitions, it also is the case that such mathematical knowledge, at some point, possesses correspondences in the physical world outside of the phenomenological checkpoint. This remains true even if the propositions of mathematics are held to be analytic *a priori*; it just becomes more paradoxical. Mathematical laws can be used to accurately predict planetary movements. How can it be that that which possesses verifiability within the phenomenological standpoint can also possess verifiability outside of the phenomenological standpoint? The answer to this question may be put into the framework of what can be referred to as analogies from experience. (This is not to be confused with Kant's use of this phrase, the meaning of which was very different and applicable only within his special system of thought.)

The first analogical level of knowledge is one which is already being relied upon whenever mathematics is employed to apply to physics. It is not at all a self-conscious reliance and if it were to be brought to the attention of a physicist, it might occasion the reaction of M. Jourdain in Molière's play *Le Bourgeois Gentilhomme*, who was surprised to discover that all of his life he had been speaking prose. Of course, it is not strictly speaking relevant to the work of the physicist to come to this awareness any more than it is necessary to know how a watch works in order to know how to tell the time. It is, nonetheless, of enormous interest to philosophers and to philosophical physicists.

There is a second analogical level of knowledge, which requires a self-conscious act. One can, in some instances, generalize from phenomenological experience to physical experience. Since this is the case, this fact can be utilized to point to a second-level analogy, which is putatively descriptive of the external world. The verifiable relationships that are found to exist between phenomenology and physics can be utilized to form a speculative set of hypotheses about what exists beyond the knowledge of the physicist. A phenomenological foundation will thus have been provided for speculation about the world that lies beyond both phenomenological and physical knowledge.

To provide a concrete example of such a general point, one may begin with the fact that a desire for knowledge of the totality or what lies beyond the limits of present knowledge is possessed by human consciousness. Such a desire, if present, cannot be totally without the possibility of being satisfied on a higher analogical plane. Just as nature does nothing in vain on the biological level, analogously, nature does nothing in vain on the philosophical level. While such analogical reasoning as this does not constitute a proof, nor is it intended as one, it is designed to carry with it a certain degree of plausibility.

No one will quarrel with the arguments that an appetite for food is nature's way of preserving the species, and sexual desire is nature's way of perpetuating the species. No one would assume that such desires would exist for no purpose whatsoever. If higher-level desires exist, it should not immediately be assumed that they could exist despite what appears as the impossibility of their fulfillment.

It would be without a foundation to begin with existential statements concerning reality as a whole. To begin with existential statements concerning reality as a whole would be similar, in certain distant respects, to begin in construction with the roof of a house. One would be starting with what normally is constructed only at the end and completes the structure. In a metaphorical sense one would also be beginning with that which is the highest component of the structure and the furthest removed from any ground. However, it is possible to begin with the fact that such a desire for such knowledge does exist just as it is possible to begin with the existence of a desire for a complete dwelling place that includes a roof over one's head. The implication that can be legitimately drawn from this is that it is possible that reality as a whole, can exist. If reality as a whole includes knowledge of that reality, then the phrase 'can exist' is preferable to the phrase 'does exist' since reality as a whole will not exist until the knowledge that it exists, exists.

Of course, it can be said that such a desire for such knowledge may exist without the equivalent possibility of such a desire for such knowledge being fulfilled. However, if the desire exists, then what cannot be denied is that the possibility of some degree of an objective fulfillment, even if in some respects unknowable, must exist. If this is denied, then the desire is a total illusion, and nature has constructed something in vain or, to put this in more modern terms, nature has constructed something which serves no utilitarian function. The desire for such knowledge cannot serve as a purely regulative function to lead one forever onwards in a quest for knowledge for if such a purely regulative function can never find satisfaction, then philosophers are doomed to a fate of eternal frustration.

In certain cases, appetites and desires possess objective means of satisfaction. There is a match between the inner desire and the outer environmental capacity to satisfy that desire. Food exists to satisfy hunger. Thus, not all desires lack objective fulfillments. There are precedents for the *possibility* that there are objective fulfillments for appetites and desires in some cases. The capacity for the satisfaction of desires in the physical realm in general may suggest that some corresponding equivalent in reality can provide for the possibility of the satisfaction of philosophical desires for knowledge of reality as a whole.

Another way of approaching this point is to consider that human beings are part of a wider system. Whatever is true of that wider system is found to be true, in some degree, of the part. Whatever partial knowledge is possessed by human beings may be utilized as a part of that system to posit knowledge claims about the whole. It may be granted that one may not be able to arrive at a complete picture of reality as a whole. However, it may be considered that reality, as a whole, must nonetheless possess the potential to exist, in order for it to be possible that one can possess that part of reality consisting of analogical knowledge which refers to that part of reality that transcends the empirically knowable domain.

If the phenomenological knowledge that is possessed, as a part, can be used to extend knowledge beyond phenomenology proper, then it can be fairly inferred, at the very least, that a whole can exist in which both phenomenological and physical knowledge are contained. Therefore, phenomenological and physical knowledge can be taken together to form an analogical bridge to posit knowledge concerning that whole. While it may not be possible to gain complete knowledge of that whole, it seems that it is undeniable that a unity between the parts must potentially exist in order for it to have been possible to have possessed a valid knowledge, both of a part and a knowledge of a correspondence between one part and another part which transcends the possible knowledge of any one part as for example with the knowledge of mathematics and the world. If a unity between the parts must exist, is it not then possible that a whole exists which cannot be reduced to an aggregate of its parts?

This is not the same as saying that because the word 'part' exists and another word 'whole' exists, one is justified in saying that a whole does exist. It is also not the same as beginning with metaphysical categories such as the categories of 'whole' and 'part'. Both of these approaches, which might, in the end, by accident, prove to be true accounts, lack any phenomenological grounding. Whether or not they lead in the end, to the same or similar results, is not important. Such linguistic and metaphysical starting points cannot make any claims upon the allegiance of the subject knower as starting points for such metaphysical starting points lack any verifiable checkpoints. They are, as Plato speaks about in the *Theatetus*, possibly right opinions, but they offer no way to systematically proceed to conclusions since their rightness (if they happen to be right) is merely accidental.

On the other hand, from a rational metaphysician's standpoint, it may be argued that the above conclusion, which one has reached, is trivial. The above conclusion that a whole may exist is a whole about which specific knowledge is lacking. This, for some, may appear to be knowledge, but the wrong amount of knowledge. For the phenomenologist proper, it may be too much; for the metaphysician it may be too little.

Can what knowledge that can be said to have been discovered now be utilized to extend further into the territory of the unknown? If a potential whole were to exist, then whatever knowledge is possessed of a part or as a part might analogically apply to the whole. A few other examples of analogical predication may now be attempted. It may be said that it has been established that the subject knower possesses as a part, certain examples of *a priori* knowledge of that part, e.g., of certain truths of mathematics. Some of these examples do have parallel instantiations in the physical world. Therefore, of certain parts, it may be said that it is at least possible that the subject knower possesses certain potential (not yet stated) *a priori* knowledge statements which may relate to reality as a whole. It has been established that apodeictic *a priori* statements exist concerning some parts already possessed. By analogical inference, it is thus *possible* that the subject knower already potentially possesses certain *a priori* knowledge of reality as a whole whether or not that knowledge is only a partial knowledge of the whole. So far, in the example that has been taken up, knowledge of the whole has been restricted to the knowledge that reality as a whole can exist. Now, to this knowledge that reality as a whole can exist, a characterization of certain parts of knowledge of reality as a whole as *a priori* knowledge can be added, and thus an additional portion of knowledge has been accrued. The foregoing conclusion is not the result of a simple analysis of the meaning of the concepts of part and whole.

So far, the results of this investigation disclose that reality as a whole can exist, and that analogical knowledge of that reality is possible. From the point of view of the present work, the whole must include knowledge of its existence in its reality. The fact is that a whole is formed which includes reality and the knowledge of that reality. In other words, there is a metaphysical reality which includes both the known and the knower within it. It would not appear that this would be to characterize reality very much to say this and thus it would still appear to be a very modest claim to knowledge. But, if one were to take even this small step, though in the eyes of some it might appear to be a mere baby step, one would thereby have committed oneself to some level of metaphysics.

Whatever knowledge is gained about reality as a whole may be utilized to assist one to better understand the knowledge already possessed concerning both phenomenological knowledge, and the knowledge that a relationship exists between that phenomenological knowledge and the knowledge of physics. At the same time, knowledge already possessed of both phenomenological knowledge, and the knowledge of the relations between that phenomenological knowledge and the knowledge of physics can act as a checkpoint on knowledge of reality as a whole. Knowledge of the whole must be coherent with, and explanatory of, knowledge of the part. Just as knowledge of the part must be a checkpoint on knowledge of the whole, so the knowledge of the whole must also refer back to and make sense of knowledge of the part. Reality as a whole must form a system. This is a conclusion that has been reached from the analysis of the relationships that already do obtain between the parts of knowledge that are already possessed. It is not a speculative starting point as a rationalist assumption. It is a conclusion that arises from an analysis of knowledge that is already possessed.

Whatever gaps in knowledge that exist do not prevent one from drawing a metaphysical conclusion concerning reality as a whole, that if a whole exists, then reality as a whole forms a system. The system that is formed is not one that entails that a deductive relationship among the parts must exist such that if one part exists that another part must exist and that all parts are parts of a chain of causation. The system that exists would be one that entailed that there was at least a partial or a family resemblance between the parts. One part may share a characteristic which is similar to another part but not every trait of one part must be shared with other parts. From this metaphysical conclusion, inferences can now be made concerning the relationship among the parts that do exist. This is not the same as saying, with Hegel, that reality is rational, or that what is rational is real, which is to begin with an unfounded premise. It is only saying that there is a systematic relationship that exists for the whole, and that the parts, which relate to each other ultimately make up a reality as a whole, which is systematic, whether or not it is rational. One's need as a human being to discover some systematic relationship that exists among the parts is thereby justifiably grounded. Whether or not such a relationship can be found is an entirely different issue. But, the very existence of the desire for such an outcome can only take its roots from the fact that such a reality already does exist in some sense. This is not a circular argument; it only demonstrates the completeness of metaphysics. What is being said is that the very desire could not exist without its being a part of a wider system; by the same token, the character of the wider system can also be employed to elucidate the characteristics of the part.

There is no question of belief being involved in this in any way. If reality were a whole, then what was claimed could not not be true. If reality were to form a whole, then the part must be in the whole, and the whole must in some degree be in the part. But, this conclusion is not a simple inference from the analysis of words or concepts; this conclusion is the conclusion to a metaphysical argument.

The progression of the metaphysical argument is to begin with the grounding which has resulted from the discovered existence of both phenomenological knowledge, and the relationship between this phenomenological knowledge and physical reality. The next stage in the argument is to consider the whole that has been consequently formed by the connection of the parts made up of phenomenological knowledge *and* physical reality. The final stage of the metaphysical argument is to consider that metaphysical knowledge has been grounded, and also at the same time provides a ground. But, it is not either grounded or a ground. As proper metaphysical knowledge it is and must be both grounded and a ground.

That reality forms a system though not a deductive system, and that therefore one can analogically infer from part to whole and from whole to part is thus far all that has been established. But, one may be able to reason further when one analyzes the particular relationships that do exist between phenomenological and physical knowledge as in the discussions of space and time. What is established here is that an *a priori* knowledge of the whole is possessed, in that an *a priori* knowledge of the whole is only possible, because of an *a priori* knowledge of the parts, and also at the same time the existence of the reality of the whole is what makes an *a priori* knowledge of the parts possible in the first place. Therefore, the existence of the reality of the whole also is the condition for the possibility of the *a priori* knowledge of the whole. If the part is the part of the whole, and one possesses an *a priori* knowledge of the part, then it follows that one possesses an *a priori* knowledge, albeit a partial knowledge, of the whole. That such an *a priori* knowledge of the whole is already possessed, although the content of such knowledge has only been partially specified, is not the same as starting with a metaphysical postulate that the whole exists. It could be said that the existence of the reality of the whole is discovered to be the transcendental condition of the *a priori* knowledge of the parts and the existence of a systematic connection between the parts is the transcendental condition of the *a priori* knowledge of the resemblance between the parts.

In the order of knowledge, the process of cognizing is from part to whole; in the metaphysical order, the order of explanation, or, the order of the conditions for the possibility of existence, is that it is the systematic nature of the whole which makes the part possible. But, what is of signal importance for this present, respective work is that one starts from the part, and not from the whole, so that one will never lose sight of the verifiable source of grounding. What makes the present, respective work unique is that the method for inferring from part to whole is the method of analogical inference, and the process is to leap from one analogical set of relationships to another, from the set of relationships that exist between one part and another part to posit characteristics that obtain of reality as a whole.

The existence of phenomenological truth is itself dependent upon ontological truth, and the analogical correspondence of that phenomenological truth with that ontological truth. The phenomenological truth appears as certain and self-evident, because it is the reflection of reality itself. What else could account for its appearance as certain and universal? The reason why what appears as true is true, or why what appears with seeming certainty and universality can possess such certainty and universality, is that appearances of phenomenological truth are microcosmic embodiments or encapsulations of a macrocosmic reality.

There is no other explanation that is possible. What have been discovered are not truths of imagination, for certain basic laws of mathematics also apply to the world. These truths of mathematics can apply to the world, if and only if, the subject knower is a part of the world, and therefore a reflection of that world, such that the laws of thought of the subject knower are also mirrored in the world. Or, to put this matter more correctly, the laws of the world are mirrored in the subject knower. The astonishing coincidence of subjective laws of mathematics and reality testing can only be explained by a metaphysical hypothesis. The intuitive necessity of the relationship that obtains in the phenomenological correlation between space and time can only be explained in the end by a reference to reality. That some metaphysical reality exists is the only explanation as to why metaphysical truths exist. Reality is the only adequate ground for appearance. There is no other way to account for necessity and universality except that what appears as necessary and universal so appears, because, it is an appearance of what is. Ontological reality becomes the ground of the appearance of metaphysical truth.

Space and Time

There has been an attempt to show the inextricable relationship that exists between space and time from the phenomenological standpoint. After such truths have been established within the phenomenological standpoint, one may legitimately consider the possibility of such truths as existing in some sense in the universe itself. If the method of analogical inference were employed, it could be said that space and time are inextricably co-involved in the universe as it exists in itself. While it is fascinating that such an implication may correspond with the theories of certain physicists, such a rationale is not the ground for the above inference. However, once more in keeping with the principle that the part must reflect the whole, it can be legitimately inferred that space and time in and of themselves must be co-defined. Again, in maintaining the principle that the part must bear some resemblance to the whole, it is to be expected that what is discovered by theoretical physics should conform as well to metaphysical truths. In maintaining the principle that the part is part of the whole, a checkpoint for metaphysical speculation remains the findings of empirical sciences such that a metaphysical picture of the world cannot be contradictory *in toto* to the findings of theoretical physics. Similarly, the whole must be instantiated in some degree in the part, or else there would be neither whole nor part and no legitimate reference to or use of these concepts could be made. This is not at all to say that metaphysical speculation must be confined to the findings of empirical science; it is only to say that it would be incoherent if no theoretical physics had some resemblance to the actual ontological structure of the universe.

In Einstein's universe, at certain speeds (or measurements of time), there is a change in mass, or space. Time and space, in Einstein's universe, are not totally separate elements. One recalls Einstein's comment on Minkowski's dictum. It is of interest to note that Einstein's theories do have a correspondence to the definitional properties of space and time on the phenomenological level, and thus provide a certain degree of evidential support that the metaphysical view of the universe presented herein may have an empirical correspondence.

The fact that a metaphysics can be shown to have a physical correlate is an occasion for the employment of a new (and different) application of Leibniz's famous phrase, *phenomena bene fundata*.[1] The ontological constitution of the world is the condition for phenomena appearing as they do. While this cannot, of course, be proved, it can be shown to have a certain plausibility. The argument for the plausibility of appearances resembling reality is that appearances or phenomenological truths possess the form that they possess in virtue of the fact that they are reflections of reality. All appearances therefore possess a kind of resemblance to reality. The resemblance of which mention is made is not necessarily a physical resemblance. If a potter were to make a pot, it could be said that the pot possesses a certain rationality, that is, a purpose. For the pot to possess this purpose, it must be that the creator of the pot is rational or purposive. It does not mean that the creator of the pot must visually resemble the pot or vice-versa. *Phenomena bene fundata* is a phrase that is now vested with a new meaning. For Leibniz, only certain phenomena were well founded; for the present work, all phenomena to be phenomena must be well founded.

Analogical predication of metaphysics is possible in the first place by virtue of the fact that phenomenological truths are reflections of reality. That phenomena are microcosmic reflections of reality is the condition for the possibility that they must bear an analogical resemblance to reality. The nature of the connection between reality and its mirror in nature is what makes analogical predication possible in the first place. This proves to be a two-way mirror, because not only does the reflection of reality in the phenomenological appearances make analogical predication of metaphysics possible, it is also the case that phenomenological reflections are *phenomena bene fundata*, and thus are reflections of reality as well as being able to reflect reality. Phenomenological truths can offer clues about reality, because phenomenological truths are themselves made possible by reality. This means that reality is the condition for the possibility of phenomenological reflections and this constitutes a transcendental argument (in Kant's sense) for the well groundedness of phenomenological reflections.

The argument that an analogical predication of metaphysics is possible is a metaphysical argument from the status of phenomenological truths as microcosms of the macrocosm. In the latter case, one infers from the phenomenological standpoint to metaphysics.

That reality possesses an analogical relationship to phenomenological truths is the condition for the possibility that phenomenological truths reflect, represent or resemble reality. The well groundedness of phenomena is therefore the reason why phenomenological truths resemble reality in the way that the child resembles the parent. The metaphysical argument is that the macro-micro link between reality and phenomenological truths is the condition for the possibility of analogical predication in the first place. The fact that analogical predication is possible also explains why phenomenological truths resemble reality, although the subject knower, has, of course, no access to reality to show the nature, accuracy or even the existence of the resemblance. The existence of *phenomena bene fundata* is thus not an argument or proof, but only a position that phenomenological truths bear some resemblance to reality, although this cannot be illustrated or proved.

Causality

It has been posited that there may be a relationship between an *a priori* knowledge of the part and the nature of the world as it is envisioned by the theoretical physicist as is illustrated by the case of space and time on one side of an analogical equation and the nature of the world as it exists in itself on the other side of the analogical equation. It may be of interest to consider another case of a possible analogical inference to the nature of reality in itself that may be drawn from the relationship that may exist between necessary philosophical truths on one level and the nature of physical reality as it is envisioned by the theoretical physicist on another level. While the case of causality is much debated, it nonetheless may be fruitful to consider what analogical inferences may be drawn concerning the nature of causality as operative in the universe itself from a comparison of a phenomenological rendition of Kant's concept of causality and the use of the concept of causality in explanations of the expanding universe as suggested by Hubble on the other hand.

While whatever knowledge one may, in the end, accrue concerning reality itself may well amount to little more than the darkest of glimpses, it will at the very least possess a degree of the foundational support that is already possessed by the knowledge that exists concerning the relationship that obtains between the *a priori* knowledge the subject knower possesses concerning the subject knower, and what is known about the physical world. Such knowledge that has been described cannot be said to be on the same level as pure speculation without any grounding at all.

To adapt causality, as a phenomenological idea, would mean that in Kant's sense, whatever idea came to consciousness after something else could only come 'after' if the subject knower related it to something else that came 'before'. From the perspective of the subject knower, the idea of causality requires the application of a rule, in Kant's sense, to correlate one event with another event that precedes it in experience. Without such a rule, there is no means of distinguishing the occurrence of an event, or something that happens, from a steady state. The description of a causal event requires the experience and contribution of the subject knower. Hence, all causal events require subjective intervention to be perceived as causal. What can one learn from this, on the second level of analogical inference, with regards to the world as a thing-in-itself?

If it is the case that causality in some sense is an unavoidable mode of thinking, and causality always requires the perspective of a subject knower, then if one were to venture a description of the world as it exists in itself, causality as it applied as a description of the world in itself would always require the standpoint of a subject knower. The inference which can be drawn from this would be that whatever causal order may obtain in the world itself would always require a subjective point of reference. This opens the door to the possibilities of the causal order of the world-in-itself to be either very similar to how it is imagined to be or extremely different from how it may be imagined to be, but not to be non-existent altogether. For Kant, one may simply not speak about (or think meaningfully about) the causal order of the world-in-itself. In the second-order analogical inferential thinking that is employed here, one may entertain the idea of causality in the world-in-itself. This corresponds more meaningfully with views concerning causality currently held by some physicists. For example, according to Hawking, since 1929 when Hubble discovered that the universe was expanding, a definite cause-and-effect relationship was established in the events of the universe. From the view of the present author, it is not clear what is the cause and what is the effect. The Big Bang would appear to be the initiating cause although its own causal status is left unexplained. What is clear is that if the universe is expanding (and now apparently at an accelerated pace), then in the past, the galaxies were closer together. There does, therefore, appear to be some causal order in that the proximity of the galaxies to each other preceded the greater distance of the galaxies to each other. However, is it not possible that this period of expansion was itself preceded by a period of contraction?

The present period of expansion is compatible with an even earlier period of expansion, which had been followed by a period of contraction. The argument of the accelerating rate would only apply to the current period of expansion but would not mitigate against the possibility of an earlier period of contraction. In the theory of analogical resemblance, one can infer that there is a causal order, but one cannot justifiably conclude that the causal order is what the subject knower thinks that it is.

All that can be justifiably concluded is that whatever idea of causal order that exists may exist from the standpoint of some subject knower or knowers whether human, sub-human, trans-human or divine. This inference is justified by its being based on the basic analogical principle that what human beings experience, as parts of the whole, must have some analogical reference to the world-in-itself. It may be said that what this amounts to is the philosophical equivalent to anthropomorphizing.

For Kant, the idea of causality must stay within the epistemological standpoint. In analogical metaphysics, it can be said that one cannot obtain complete knowledge of the causal order of the universe, because in the universe itself, every causal order also depends upon some epistemological standpoint. If the causal order of the universe always must depend upon some epistemological standpoint, ontology always depends upon epistemology for its discovery. This is not the same as saying that one can have no knowledge of the thing (or the universe) as it exists in itself. It is saying that there is no such thing as the thing-in-itself, if by that is meant a metaphysical order which is entirely divorced from epistemological standpoints.

In the end, the perspective introduced in this work does not reduce to a perspective in which all that can be known is within the viewpoint of some subject knower such that every *esse est percipi aut percipere* in the fashion of Berkeley. In the present approach, every *esse* is in part a *percipi*, or to be more precise, a *percipere*, but this forms only a part of the *esse*, and not the whole of it. The system which the universe forms (not which forms the universe) is a system which may involve epistemological *and* metaphysical elements which relate in a chain in an analogous fashion to the way in which Aristotle viewed the ontological levels of causation. For Aristotle, each ontological level was related to another ontological level. From the point of view presented here, every ontological level is related to some epistemological level and vice-versa, such that epistemology and ontology are inextricably co-involved in any descriptive account of the universe. It is not that the universe is unavailable to the subject knower because of the interference of epistemology; it is that the universe is inherently composed of both epistemological and ontological elements. For Heisenberg, the world of physics is uncertain in itself and requires human intervention to bring it to some level of determinateness. In the present view, the level of determinateness of the world is the result of a co-production of many epistemological standpoints. It is as if the condition described by Heisenberg were actually true of the universe itself.[2] This is not because of the difficulties involved in physical measurements of the universe; it is because it is characteristic of the universe in and of itself to be an epistemological and a metaphysical blend. But, this is not simply a speculative metaphysical claim about the universe; it is grounded on the phenomenological experience, which characterizes human knowledge, and is an analogical inference from that knowledge condition.

What implications follow from the above? The first implication is that there is a causal order of the universe (although this causal order may be seen differently from differing epistemological reference points). It cannot be that there is no causal order to the universe, since in human experience there is some causal order, although what the causal order is differs in its description according to different subject knowers. What the causal order of the universe is, however, may in the end, be very different from what it is imagined to be. There is no reason why the causal order of the universe would have to resemble the causal order according to the human reference point. This opens up the possibility that the causal order of the universe may in fact be very different from various human imaginings of it. This is, however, very different from saying that there is no causal order in the universe at all, or that whatever causal order that exists in the universe is that which is forever unknowable to the human subject knower.

Descartes' Causal Proof

The causal proof of Descartes' is, in a way, the most primary ontological argument. Anselm's argument, that if one were able to conceive of the most perfect being, he must exist, is still dependent upon Descartes' causal argument. The very ability to think of the perfect being in the first place is dependent upon the reality of that perfect being. Descartes' causal ontological argument is in fact the most primordial of the ontological arguments.

A typical rejoinder is that one can arrive at the idea of perfection simply by the process of adding increments of "betterness" to everyday imperfect experience. The idea of perfection thus would not carry any definite meaning, and therefore would not require any Divine origin. This will probably be debated between absolutists and empiricists forever. But the arguments advanced here concerning the sophisticated whole-part relationship (described in greater detail above) would appear to be on more stable ground.

To sum up. The starting point is the existence of phenomenological knowledge. From such knowledge one can analogize to a world outside of phenomenological knowledge. Since the physical world possesses relationships in it that bear analogical resemblance to corresponding relationships that obtain within phenomenological knowledge, it appears as if the phenomenological world can serve as a clue to the nature of the outside world. This is no small discovery since it offers grounds that consciousness can serve as a reliable, analogical guide to reality. What is more, consciousness could not serve as such a clue unless the outside world first existed, since clues are left by that of which the clues are clues. It would make no sense for the clue to exist without the existence of that of which it is a clue. It is that which exists which leaves the clue, and makes the clue possible in the first place. This latter argument serves as a proof of the existence of reality as a transcendental condition for the existence of the knowledge of reality.

While it may not be the case that the Cartesian causal proof is a wholly reliable source of the proof of the existence of perfection - it may be the case that the possibility of analogical predication of reality can exist in the first place if and only if there is a reality which is being analogized. An analogy can only exist if the terms to be related exist. Thus, if it were possible to say that the phenomenological world serves as a clue to the relations that exist in the physical world, it may be the case that the possibility of the existence of the analogy is based on and given by the ontological existence of this world structure (including both potential phenomenological and physical worlds within it). This world structure is a reality, which is capable of being known at least as the causal condition or cause of the existence of both phenomenological and analogical knowledge in the first place. This may be the first valid context for the application of Descartes' causal ontological proof: a proof not of the existence of perfection, but of the existence of ontological reality.

In a similar vein to Descartes, the methodology of the present work is not to commence with the idea of G-d, or in this case, ontological reality, but to commence with the possibility of phenomenological truth. The next step in the methodological procedure is to conceive of the possibility of analogical predication. The final step in the methodological procedure is to infer that the reality of an ontological structure of the world is the condition of these dual possibilities. According to the present doctrine of *phenomena bene fundata*, phenomenological truths are true and well-grounded phenomena, because they are reflections of embedded metaphysical truths. This accounts for why it is possible to analogize from space and time on the phenomenological level to metaphysics, and why it is possible to analogize from the truths of physics to the truths of metaphysics.

Truth

The existence of phenomenological truth is itself dependent upon the existence of ontological truth, and the analogous correspondence of that phenomenological truth with that ontological truth. Unlike older metaphysics, the major question with which this present work is concerned on the level of the general theory of metaphysics is not whether or how the empirical world resembles reality, but rather why certain phenomenological truths appear as self-evident, or, what makes it possible that certain propositions have the character of appearing as apodeictically certain truths? The resolution is that certain propositions appear as self-evident, because they report truths that are reflections of reality. In other words, propositions appear as true, because they are reflections of a true state of affairs. One can understand something to be true if and only if it is true: the condition of a subjective apprehension of truth is its grounding in objective reality. Otherwise, the possibility of what is true appearing as true is unintelligible.

Certain truths appear as self-evident because they are the reflective appearances of reality, just as a mirror reflects what is there, because there is something there to be reflected in it. The truth-appearing character of propositions cannot be explained on the grounds of the individual propositions themselves, because that would be question begging. That some state of affairs exists in reality is the explanation as to why that state of affairs would appear to exist of necessity. Ontological existence is the ground of metaphysical truth appearances. This is a secular equivalent of the Cartesian causal proof of the existence of the Deity.

It could, of course, be argued that any proposition appearing as a metaphysical truth is only true because of a matter of definitions. But, this will not explain why certain metaphysically truth-appearing statements are also true of the world, such as certain mathematical propositions. It is only by hypothesizing the unified structure of reality that the fact that certain mathematical propositions both have a subjective truth-appearing characteristic, and also apply to the world, can be made comprehensible.

It seems that the only feasible explanation for why human agents can understand that which is certain and which also appears to be true of the world is that the rational faculty is that level of consciousness which is the self-consciousness of the world. *It is only if the subject knower is a part of the world, and therefore a reflector and a reflection of that world, that it is possible that the laws of the world are mirrored in subjective consciousness.*[3] Or, it may also be said that the laws of consciousness are mirrored in the world. The astonishing coincidence of some forms of subjective mathematics and reality can only be explained by a metaphysical hypothesis, to wit, that consciousness is a part of the world and the knowing subject is both a reflecting consciousness and a conscious reflection of that world.

Metaphysical reality is not, as Kant thought, unknowable. Just as mathematical laws are imaged physically in the world, a correlation must exist between metaphysical laws and reality. Just as mathematical laws are not literally repeated in the physical world but are exemplified in it, metaphysical laws are characterized by and characteristic of reality. This is another way of saying that truth can be described as an analogical adequation with reality.

Toward a New Metaphysics

The kind of metaphysics that can be conducted today is not pure speculation. It is grounded on phenomenological truth discoveries and is analogously related to those discoveries. Kant thought that the subject knower had no knowledge of things-in-themselves. But, while it may be true that the thing as it exists in itself is unavailable to the subject knower directly, it does not follow that there is no possibility of an indirect or analogous knowledge of the thing-in-itself. Analogical predication, has, of course an honorable history in Medieval Western philosophy.

From the standpoint of the present work, the first step is to establish some phenomenological truth, which is grounded in the immediate givenness of consciousness. The present work does not begin with a pre-existent assumption, e.g., that the Deity must be omnipotent. The present work begins with phenomenological truths that cannot be gainsaid. In the present work, it is inferred that truths can only exist as truths, if and only if such truths are particular illustrations of an existing true state of affairs. For example, in noting that there is a relationship between the subjective truths of mathematics, which prove to have objective referents, it is to be inferred that an analogous relationship between discovered metaphysical truths and reality might exist. But, the distinction between the use of analogical predication here, and one of its uses in the Western Middle Ages, is that in the present work much attention is paid to the starting point which is with *discovered* and not with *given* or *revealed* truths.

If one can arrive at some version of metaphysical knowledge of reality, then the subject knower has access to at least an analogical knowledge of things-in-themselves. And, the analogical of knowledge of things-in-themselves possesses a regulative use as well. Such knowledge may prove useful in guiding the direction of scientific inquiry, and should also be reflected in truths of science. For such metaphysical discoveries have a two-way direction. After discovery, they can also be used backwards to aid the course of scientific inquiry. If man's relationship with mathematics is also grounded in the physical world, and the understanding of how this is possible constitutes a metaphysical act of understanding, and that understanding, and that of which it is an understanding, cannot exist unless it is a reflection of and a revelation of reality, one thereby possesses an indirect understanding of things-in-themselves.

In the end, the present work can be described not only in the manner of Paton's understanding of Kant's thought as a metaphysics of experience, but, as a metaphysics of transcendence. One arrives at an understanding of what reality must be like in order to provide an explanation for the possibility of understanding the relationship between man and the world. At the same time, one understands that it is the existence of reality that makes it possible that the relationship and the knowledge of that relationship can exist in the first place. A metaphysics of transcendence is a valid body of knowledge, and must exist as a valid body of knowledge in order that there be a metaphysics of immanence, or a metaphysics of experience. Ontological truths must exist in order that there can be phenomenological truths. The ontological truths are the grounds for the existence of the epistemological truths, and the epistemological truths are the gateways through which one comes to understand the existence of the ontological truths. Ontology is the ground of epistemology, and epistemology is the discoverer of ontology.

Notes

1. In the present work, the sense of *phenomena bene fundata* is different than Leibniz's sense in that the usage of the phrase *phenomena bene fundata* in the present work refers to *phenomena bene fundata* as clues to the nature of reality whereas for Leibniz, the phrase *phenomena bene fundata* gave information about the nature of the external world. *Cf.*, Benson Mates, *The Philosophy of Leibniz, Metaphysics and Language*, New York: Oxford University Press, 1986, p. 198. There is another difference in that *phenomena bene fundata* in the present work are well founded when they are founded by metaphysical reality. For Leibniz, *phenomena bene fundata* were well founded because they were accurate reports of the external world and thus were epistemologically and not metaphysically grounded.
2. W. Heisenberg, Z. *Physik*, 43, 172 (1927). While, in the present author's understanding of Heisenberg, according to Heisenberg, the path of an object first comes into existence when it is observed and thus uncertainty is characteristic of the universe itself, Heisenberg's reasoning as to why this is the case seems to be based on difficulties arising from the problematic nature of observation. Aristotle, in contrast, provides a metaphysical explanation for why events in the universe can never be fully predictable and thus the universe is not determined. For Aristotle, spontaneity is an inherent characteristic of the universe and is a wider category than chance. *Cf.*, *Physics*, Book II, Chapter 6. Kragh finds a positivistic quality in Heisenberg when he quotes Heisenberg as saying, 'If one wants to make clear what is meant by the words 'position of an object,' for example of an electron ... then one has to describe definite experiments by means of which the 'position of an electron' can be measured; otherwise this term has no meaning at all.' For Kragh, Heisenberg's 1927 paper meant that the classical concept of causality had to be abandoned because a physical system could never be defined precisely. On the other hand, as Kragh points out, uncertainty relations do not rule out causality. *Cf.*, Helge Kragh, *Quantum Generations, A History of Physics in the Twentieth Century*, Princeton: Princeton University Press, 1999, pp. 208-209. The cyclical universe is discussed by various physicists. *Cf.*, Helge Kragh, *Cosmology and Controversy*, Princeton: Princeton University Press, 1996.
3. It is possible that this provides the clue to a solution to the problem which Kant raises in his February 21, 1772 letter to Marcus Herz in which he writes, 'I noticed that I still lacked something essential, something that in my long metaphysical studies I, as well as others, had failed to pay attention to and that, in fact, constitutes the key to the whole secret of hitherto still obscure metaphysics. I asked myself: What is the ground of the relation of that in us which we call "representation" to the object?'? And later, he states, ' ... this question, of how the faculty of the understanding achieves this conformity with the things themselves, is still left in a state of obscurity'. *Cf.*, *Kant, Philosophical Correspondence 1759-*

99, Arnulf Zweig (ed. and trans.), Chicago: The University of Chicago Press, 1967, pp. 71-2.

Largo profundo
Slowly and with depth

8 A Metaphysics of Virtual Reality

The Inner and the Outer Circle

In Chapter Four, it appeared as if space was argued to be a transcendental condition for the possibility of time and vice-versa. The arguments in Chapter Four were metaphysical arguments in the sense that they disclosed the concepts of space and time as *a priori*. In this regard they were metaphysical expositions in the Kantian sense. The metaphysical arguments for space as the condition for time and vice-versa that have been offered are different from Kant's metaphysical expositions, which were exclusively directed at showing that space and time were concepts that could be exhibited as *a priori* concepts. In Chapter Four, space as an *a priori* concept was shown to be *a priori* for time and vice versa. For example, it was shown that successiveness cannot be illustrated without representing the succession as being in the same place. Without the constancy of place, one thing cannot be represented as replacing another. While for Kant arguments were offered for space and time as transcendental conditions for the possibility of mathematics, arguments in Chapter Four were offered for space and time as transcendental conditions for the possibility of each other. While strictly speaking this may appear to be arguing in a circle since it has also been argued above that there is no such thing as space and/or time by themselves, showing that space and time are transcendental conditions for each other is another way of illustrating their interconnectiveness; it is not meant that they are literally conditions for each other as if one could exist without the other as separate entities; this would be to confuse a transcendental argument with an argument for physical conditions. The arguments offered in

Chapter Four should best be characterized as metaphysical arguments for the equiprimordiality of space and time.

However, the above argument does possess all the appearances of a *petitio*. Since space and time are in reality the same concept but viewed from different aspects, each concept is both the transcendental condition for the other and at the same time is *a priori* only if and when the other is *a priori* so that they are equally metaphysically intertwined. In other words, the metaphysical argument for the condition for the existence of the one cannot be made without at once demonstrating the other as equally *a priori*. But this is not a mere metaphysical exposition because it is only if one concept exists as *a priori* that the other can exist as *a priori*. This is the nature of a transcendental exposition and not merely a metaphysical exposition. In the transcendental exposition, the one is shown to be impossible without the representation of the other. In the metaphysical exposition, as the phrase is utilized in this present, respective work, the demonstration of the *a priori* character of the one cannot be made without at once displaying the *a priori* character of the other. In the case of space and time, the transcendental exposition and the metaphysical exposition are one and the same.

While it appears as if space and time are transcendental conditions for each other, this cannot be the case. They are conceptually correlative, and one cannot exist without the other. On the other hand, they are not the same concept. They are genetically twins but they are not identical twins. It is the case, however, that one cannot present the *a priori* nature of space without at the same time displaying the *a priori* nature of time. Thus, they appear as transcendental conditions for each other, but this is only a feature of their interconnectiveness. One cannot show the *a priori* nature of the one, that is, offer a metaphysical exposition of the one, without simultaneously showing the *a priori* character of the other. In reality, there are not two separate concepts, one of space and the other of time, but only a concept, space-time.

Since empirical conditions are simply the empirical reflection of metaphysical and transcendental conditions, the metaphysical connection and the transcendental connection of space and time are reinforced by empirical discoveries. The metaphysical character of the argument is that the *a priori* character of space is indissolubly connected with the *a priori* character of time. The transcendental character of the argument is that they are not only indissolubly connected but each is the transcendental condition of the possibility of the other. It becomes plausible that one can analogically suggest that space and time are ontologically coexistent. While this cannot be demonstrated with the same force or to the same degree as the demonstration that space and time are either transcendental or metaphysical conditions for each other, nonetheless it becomes appropriate to suggest an ontological parallel for the sake of completeness. Such an ontological parallel was already pointed to in Chapter Seven. Here, there is occasion to elaborate further implications of the ontological parallel.

Space and Time as Transcendental Conditions for the Possibility of Representation in General

Language is spatial whether it is ideographic, pictographic, alphabetical or the language of binary numbers. Words or numbers or symbols in general (hereafter the word 'word' or 'language' will be used to stand for all symbolic forms of representation whether one is referring to the language of musical notes or microscopic marks on a computer chip) must be extended in space whether in micro or macro space. In addition, language is temporal in that each word unit must replace each other word unit in succession. Space and time are transcendental conditions for the appearance of language.

It has been argued that it is impossible to think without the use of language. This however is not strictly speaking an accurate rendition of the epistemological moment. Strictly speaking, the idea of 'thinking' is already linguistically committed. The entire representation of 'thought' as a discursive process is an illusion. Thought never takes place discursively. All discursive thought is already temporally projected as successive and spatially projected as extended. For the sake of convenience, space may be selected for special emphasis. Thus, it can be said that discursivity is the spatial form of the representation of the thought process. *No one in the history of humanity has ever possessed a discursive thought.* Discursiveness is simply the *ex post facto* means by which thought is represented. The existence of the concept of discursive thought merely demonstrates that spatiality is unavoidable as a means of representation. In fact, the existence of discursivity is a proof of the primordiality of space.

It is wrong to attempt to avoid the implications of the unavoidability of spatial forms of representation. It is more interesting to note that while it is possible to transcendentally isolate an idea from its mode of representation (the fact that the two can be thought as separate concepts implies that they are logically separable although it does not imply that they are ontologically separable), they cannot be so represented. That they can be logically separable shows that thought *qua* thought does not require language and vice-versa (speaking here of the moment of thought and not thought in general and the moment of language employed and not language in general). But that they are logically separable is distinct from the fact that they cannot be empirically represented as separable - hence the common saying and belief that thought cannot exist unless it can be put into words.

If one does not discount the necessary spatial forms of representation (common to all language systems) but rather takes special note of them, one can become more familiar with the truth that in order for there to be common knowledge, there must be at least two sources of knowledge. First of all, there is the knowledge of pure ideation; secondly, there is the knowledge of the form of pluralization, which must always accompany and is the necessary form of the representation of ideation. Conceptual knowledge, strictly speaking, does not exist at all but ironically is frequently perceived of as the only kind of knowledge that does exist.

While a more complete account of knowledge is to be found in Part II, Chapter 1 of the present author's *A Metaphysics for the Future*, a more simplified version is offered here. The awareness that there are two sources of knowledge is a transcendental reflection. The term 'intuition' is to be avoided unless it is introduced with appropriate qualifiers. The two sources of knowledge (*zwei Erkenntnisquellen*), which Kant remarks upon *en passant* in his first *Critique* but seems not to further develop, are not sensibility and understanding by concepts as he normally taught.[1] In the work of the present author, one source of knowledge is non-conceptual or the knowledge of pure ideation; the other source is found only in employment and via reflection on the means of representation and is the knowledge of pure difference. The difference between the two kinds of knowledge has nothing to do with spontaneity and receptivity. The difference between the two sources of knowledge is that one operates, as Plato discussed in *Republic*, by means of pure ideas through pure ideas to pure ideas and the other operates by means of what is not an idea but without which no idea can be represented: the pure form of difference.[2]

The second source of knowledge cannot strictly speaking be thought at all: it can only be represented through the forms of space and time. *A word is a fusion of the two sources of knowledge*. A word can be analyzed into two components, meaning and sound (or visual representation). It is impossible to refer to an idea or a language unit without meaning and sound (or visual or other sensory depiction). The common concept that thought cannot exist without language is a distorted representation of reality. In reality, thought and language must exist separately as epistemological realities because of the simple fact that while language must be spatial and temporal, thought is not bound by such parameters although it cannot be represented otherwise.

Any simple word taken by itself shows the unity of thought and difference. Pure difference (not a difference in idea) can only be represented in space or time. Pure difference itself cannot be cognized. Pure difference is not an idea. Pure difference can only be shown through spatial and temporal ostension. Every word, every idea, has two sides. Every word and every idea must be represented in space and time but one aspect can only be thought and the other aspect can only be represented. Non-discursive thought and representation are two different sources of knowledge. These two sources of knowledge have nothing to do with spontaneity of concepts and receptivity. Spontaneity and receptivity are empirical conditions and possess no place in a transcendental inquiry. *An idea is a specious notion.* There is no such thing as a concept. A concept is already a mixture of difference and pure ideation. Strictly speaking, there is no such thing as an idea which exists by itself.

It is best, perhaps, to begin with the unit of the word. A word can be analyzed into its cognitive meaning and its differentiating function. Though a word, strictly speaking, is a fiction, it is a fiction which, when analyzed, is less liable to lead one to greater falsifications than if one attempted to analyze the idea of an idea. In terms of degrees of falsity, the concept of a word is less false than the concept of an idea. Pure difference is also a fit subject for transcendental investigation. Words and instantiations of pure difference, which possesses spatial and temporal attributes, can be investigated more easily than the pure idea, which possesses no differentiation. Indeed, the expression 'the pure idea' is but a convenient fiction. But it is a necessary fiction. For human inquiry, more substantial results can be obtained if one focuses attention on the forms of difference (space and time) and the difference between these forms of difference and the pure idea (which can also be of difference), rather than on the pure idea itself. But these topics are treated in the present author's *A Metaphysics for the Future*.

Application

Of what application are the arguments concerning space and time to the Great Questions of philosophers such as, was there a beginning to time and was there a beginning of the universe? While such questions have already been touched upon, the discussion can now be enlarged and taken in a different direction. With respect to certain physicists, the universe does have an origin in time. But some philosophers have possessed difficulty with this concept since the question can always seemingly be raised, what happened before that? Since the question can always be raised, does that not imply that there could have been a time before time existed? The very concept of time would seem to imply an infinity of itself. Therefore, the universe could not have had a beginning.

The problem of the beginning of time constitutes a most startling paradox. It seems to be a case in which language and concepts overrun reality. For Descartes, error was due to will overcoming reason. Here, it seems as if human beings are prone to error because reason overruns reality. Occasionally, there is more in philosophy than exists in Heaven and Earth.

The fact that it can be *thought* that time can always have a beginning or that the nature of time is to be infinite or to contain an infinity inside itself or outside of itself does not imply that physical time must incorporate these features. While there is an analogy between the physical universe and conceptualization, there does not need to be an absolute identity. In Chapter Five, a suggestion was made to account for the inability to solve the puzzle of the paradox of the beginnings of space and time. In Chapter Five, the special metaphysical postulate of the inter-ontological connection of space and time was proposed as a reason for the insolubility of the problem. The inability of the human mind to explain the origin of time in space or space in time was a reflection of the nature of the universe. Here, the seeming conundrum created by proposing a temporal origin of space while at the same time proposing a spatial origin of time is considered as arising from a confusion between the nature of phenomenological space and time and physical space and time. The two approaches are compatible in that the former approach explains why the problem exists while the present approach explains how the problem manifests itself and how it may be avoided.

Why time may appear as absolutely infinite is due to the nature of possibility. But actuality is always limited. While itself limited, actuality is also a limit: actuality is the limit of possibility. In order that it is even possible to represent the potential infinite, there must be a representation of a finite limit. Potential infinity can be represented only if an actual finite can be represented to which one can add increments *ad indefinitem*. If there is no representation of an actual finite, that is to say, no limit, how can an indefinite expansion beyond the limit be represented? The representation of the finite is the condition for the possibility of the representation of the potential infinite. It is true that as soon as one draws a boundary that one is outside the boundary. But the boundary must be represented first before one can transcend it. That one is already beyond the limit in order to draw the limit is due to the nature of phenomenological space which makes representation possible in the first place. In terms of the construction of concepts, it cannot be said that it is the representation of the potential infinite that is the condition for the representation of the actual finite. Anselm and Descartes thought that it was possible to conceive of the absolutely infinite or perfection. For Descartes, it was the existence of the perfect that made it possible to conceive of the imperfect. But it is difficult to grant that one may conceive of absolute perfection or an actual absolute infinity. It is possible, however, to conceive of the finite. If it is possible to represent the finite, it may be that an actually existing finite can be inferred. If finitude were a matter of pure possibility, how would it be possible to distinguish the infinite from the finite? Finitude cannot be a matter of pure possibility or how or why would the concept of the finite even arise? In order to fulfill the condition of an analogical adequation of representation with reality, on a higher level, could it then be inferred that there is a limit to reality?

The microcosm must bear some resemblance to the macrocosm. If it were to be said that the capacity for representing the potential infinite implied that the universe were infinite, this would be to ignore how the concepts of finitude and infinitude were constructed. The condition for the possibility of the construction of the concept of the potential infinite is the representation of the finite, not vice-versa.

Since a finite span of time may be represented, it would appear to follow from the above argument that, in reality, time possesses a finite dimension. If time is limited, then what limits time? Time cannot be limited by itself. If time were limited by itself, as Kant foresaw, one would be involved in a self-contradiction. Time, limited by itself, would be limitless and hence could not limit itself. But time, by the argument above, must be limited. What limits time is not itself, but another. The limitation of time is space.

With respect to space, a similar dialectic is operative. One can think of space as limitless but it does not follow that physical space is limitless. Kant thought that because to think about either space or time involved oneself in contradictions, one could not with the use of pure reason come to any conclusions as regards the nature of reality, as to whether it was limitless in space or had a beginning in time.

Kant's own arguments, however, prove something else. By starting with the thesis of the [empirical] finiteness of space and time, he proves their [phenomenological] boundlessness [in one's imagination]. While his conclusion is completely correct as far as phenomenological space and time are concerned, in his argument on both sides, both in his proof of the thesis and his proof of the antithesis, he commits a transcendental misapplication of his categories. He raises questions concerning the nature of space and time as things in themselves. While ostensibly physical space and time are the objects of his inquiry, the real objects of his arguments are [or should be] phenomenological space and time. When he begins with the concept of space and time as being finite, he is at that moment discussing the nature of physical space and time. The thesis of the finiteness of space and time is not a thesis that applies to phenomenological space and time.

The conclusions that he reaches as regards physical space and time as having to be limited with respect to his proof of his thesis, do not in any way contradict the nature of phenomenological space and time. The fact that physical space and time are limited is not affected by the fact that one can think of a space outside of space or a time outside of time. Phenomenological space and time always exceed physical space and time just as our thought exceeds reality. One can imagine oneself as immortal but physical reality cannot match this imagination.

In fact, Kant has also mixed up the methods of empiricism and rationalism with those of phenomenology. In phenomenological imagination, one can always imagine more space or more time than may exist in physical reality. The Antithesis, (the portion of it that is relevant) that space and time are infinite, better worded is not a thesis (for that implies a rational hypothesis), but a phenomenological fact discovered in an active process of imaginative experimentation.

Kant thought, of course, that he was comparing two rational hypotheses. He was not aware he was comparing a rationalistic one [that space and time are finite] with a phenomenological one [that one can always imagine boundlessness within and without]. Nonetheless, his argumentative methodology was sometimes phenomenological in nature as in the Transcendental Aesthetic, although he did not label it as such as he was not seemingly aware of it as such. Kant also was not aware of the interconnected phenomenological nature of space and time. When one takes into account the phenomenological nature of space-time and the difference between phenomenological space-time and physical space-time, one is in a better position to approach a solution to the problems Kant couches in his First Antinomy.

That one is capable of phenomenologically projecting infinity only proves that possibility exceeds actuality. Possibility is unlimited; actuality is limited. One could not imagine an infinity outside of or inside of space or time unless it were possible to represent an actual limitation that made it possible to indefinitely add an additional increment and indefinitely subdivide an existing unit. The only limitation to time, since it cannot be time itself, must be space. The limit to the endless subdivision of time is the amount of space available. The limit to the endless process of time is again the finiteness of space. The limitation of space is the amount of time available. Space cannot expand without the availability of time.

The absolute beginning of time is space. The absolute beginning of space is time. In a sense of course this is arguing in a circle since space and time possess no existence in themselves. But time, as the number of motion, requires space to number. Time begins with the emergence of space. The time of the universe begins with the origin of the universe. In this respect the views of certain physicists are correct.

The end of the universe in time can be calculated and has already been so calculated by certain physicists. The universe will end in time with its self-destruction in space. The two events must be concurrent. Time ceases and can only cease when there is no space for it to occupy. Space ceases and can only not be when there is no time for it to occupy. Space only exists in time and time only exists in space.

Kant and Metaphysics and a Metaphysics of Virtual Reality

Kant was right to criticize the powers of pure reason to know reality, but he was right for the wrong reason. It was not that human beings have no access to the thing-in-itself that makes metaphysics in the rationalistic sense impossible; but that human beings possess no direct access to the thing-in-itself provides a clue to the nature of reality. On the grounds of the principle of the microcosm resembling the macrocosm, it is likely that reality will itself possess the feature of being a mix of epistemological and ontological levels. This is not a principle that is 'read into' experience because it is only if phenomenological truth is first experienced in its necessity that an ontological ground for that proof can be sought. In order to do metaphysics, one must begin with phenomenological data. These data are pure objects of consciousness. Among these data are certain data that reveal universal and necessary truths. It is from this subset of phenomenological data, the data of universal and necessary truths that one makes analogical inferences concerning the nature of physical and metaphysical reality. The inferences which one draws do not constitute, as the claims of the rationalists of the seventeenth century, direct knowledge of reality. They constitute, descriptions of what reality is nearly like, or, as it were, descriptions of virtual reality. It is not necessary, of course, to make inferences concerning the nature of physical reality since this may be adequately, nay, even more adequately be taken care of by the physicist. But one may nevertheless simply point to directions which the physicists may not yet have considered taking.

With regard to the special metaphysical structure of the universe, conclusions may be inferred that are not phenomenological inferences. They appear to be rationalistic inferences. However, they differ from the assumptions and inferences of rationalist metaphysics in two key respects. First of all, they are inferences which are not assumptions or based on assumptions but are inferences which take their rise from phenomenological grounds. Secondly, they are inferences which do not stake their truth claims on the grounds of certainty and universality but stake out the more modest claim of analogical likelihood. Inferences concerning the universe assume the form that, 'it is likely that the universe is constructed in such and such a fashion', or, 'the universe is most likely to resemble this truth in some approximation if not exactly identical to the descriptive claim that is being made about it'. This is a metaphysics of virtual reality.

For Kant, one could not aspire to metaphysics because one was always limited to appearances - to the way the world appeared to human beings. Herein, the human being, as a part of the universe, must partake of the quality of the universe, as the microcosm is a part of the macrocosm. The human being, being an intelligent consciousness, can understand the workings of the universe. This capacity to understand has its roots in the universe itself. What is intelligible to the human being is intelligible because the object of knowledge - the universe - is akin to the powers that know it. One can possess certain knowledge of truths and analogical knowledge of that which makes the truths possible and that which makes the truths true, namely, the universe itself.

For Kant, pure reason can never know reality. Here, phenomenologically based reason, that is, reason which bases itself upon the subset of necessary and universal data of phenomenology can both possess the certainty and universal truths of the phenomenological data themselves and at the same time bear analogical truths of reality. Metaphysics is thus grounded on phenomenologically based reason and when so founded can keep itself from degenerating into pure speculation and fantasy.

One must bear in mind that Kant's critique of pure rationalism is correct, though not for the reasons that he thought it to be correct. Pure rationalism is inadequate because thought requires an anchor. Kant's anchor, that of empiricism, is the wrong anchor. The anchor that is required is a phenomenological anchor. The phenomenological anchor can return to Kant the necessary and universal truths which he himself thought to exist; it is only that they are, for the most part, a different set of truths; they are known to be necessary and universal for different reasons, and they serve an altogether different purpose.

The situation of present-day phenomenology is that, most ironically, some phenomenologists have concerned themselves with the realm of empirical appearances, which possesses the dubious distinction of being the least interesting and the most plundered territory of the phenomenological and empirical inquirer alike. The inquiries of some present-day phenomenologists into culture and society bear such a resemblance to the inquiries of sociologists or cultural critics that it is difficult to detect what is phenomenologically unique in them. Few follow in the footsteps of the early Husserl. To the phenomenological inquirer, the territory of necessary and universal truth is by far the most interesting terrain and the one which has been the least explored. It is from the boundaries of this least explored territory that one can posit analogical truths concerning the universe, which are of the greatest interest to philosophers, physicists and ordinary human beings alike.

Notes

1. B29.
2. Plato, *Republic*, VI. 511; *Parmenides*, 146D.

Andantino risoluto
Slowly and resolutely

9 On Fact and Essence, the Philosopher and the World

On the Inseparability of Fact and Essence

In Part I, Chapter 1, Section 2 of *Ideas*, Husserl raises the intriguing topic of the inseparability of fact and essence. Quite apart from what Husserl may have meant by his brilliant discussion, e.g., contingency possesses an essential character, from the point of view of the present, respective work, the existent actuality of fact is a pure contingency. This may indeed be what Husserl means when he says that, '... every fact could be "essentially" other than it is'. But the discussion of this chapter takes its rise from being stimulated by the raising of this topic and is not necessarily devoted to following Husserl's direction.

From the point of view of this present, respective work, there is no necessity that any existent fact (apart from its being known as a 'fact') requires the existence of essence. By fact what is meant here is simply contingent existence. If one meant by fact a state of affairs which was known to be 'a fact', of course a scientific theory of truth would already be assumed and for such a factual existence, essence in the sense of some degree of lawfulness in the universe would already be presupposed. However, in the sense of fact employed here, in the sense of pure facticity apart from any knowledge claim, there is no requirement of essence.

Nevertheless, there is another sense in which there is an inseparability of fact and essence and that is in the domain of essential facts. If one speaks of an essential fact one is referring to a fact which is a datum of necessary knowledge. When one speaks of mathematical truths of a certain variety, one may refer to such truths as essential facts. Such essential facts possess the status both of being necessary truths and of being part of the structuredness of the universe and thus as being metaphysically necessary. However, even such essential facts could not exist. The universe itself could not exist.

If essence is unknown, then it cannot be said that there is a fact of essence. An essence cannot become a fact unless it is known and then and only then is it proper to speak of the fact of essence. Hence, there is no inseparability of fact and essence when one refers to essential facts. It is only that once an essence is known, it becomes a fact of essence.

The state of facticity is pure contingency. There is no necessity that contingent facticity exists. With respect to essence, there is no necessity of the existence of fact. Essence *qua* essence does not require facticity for its existence. It may be that such an essence would be unknown and unknowable, but it does not follow that thereby it would be void of reality. It might be that it would be impossible for it to be comprehensible as existing, but this only means that the category of existence is inappropriate for its description. Facticity, with respect to its existence, does not require essence for its being; essence *qua* essence does not require facticity for its being. Facticity and essence, therefore, are independent from each other, although not epistemologically separable.

Building a Bridge to the World

An interesting route through epistemology can been constructed by which one can safely travel to the destination of metaphysics. It is the duty of the philosopher not to shirk from recognizing the implications of her or his discoveries. In the process of knowing truths, one can become aware of the distinction between different kinds of truths, as for example, between necessary truths and contingent truths. Further, one can become aware of the apodeicticity and universality of the distinction between necessary and contingent truths. The question then arises, when one becomes aware of the distinction between necessity and contingency, does it follow that such a distinction would not have been possible unless one were already aware of contingency and therefore that one's awareness of the essential nature of truth would not have been possible if one did not at first possess the ingredient of the contingent nature of truth, an ingredient which could not have arisen from the realm of essence at all? In other words, in order to have become aware of necessity as necessity one would have to have been aware of contingency. Thus, an awareness of necessity necessitates an awareness of contingency. But contingency could not have arisen from the realm of necessity. If contingency were already part of that which was necessary, it would not be contingent. Therefore, it would be necessary that contingency exist in its own separate, ontological realm. It would not follow that for necessary existence to exist, contingent existence would have to exist. It would only follow that for one to become aware of necessary existence, contingent existence would have to exist.

The question can then be raised here, given the starting point of pure consciousness: can one know only the necessary truths of essence? Whether or not the world exists is outside of the realm of necessary truth and thus outside of the scope of the concern of the phenomenologist proper. It does not follow from this that the world either exists or does not exist; it merely follows that it is not a subject about which a pure phenomenologist *qua* phenomenologist has anything to say. The question nonetheless remains, even if it were to be the case that the philosopher, *qua* pure phenomenologist, possesses no interest in proving the existence of the world, can the philosopher *qua* metaphysician or epistemologist contemplate essence *qua* essence without presupposing the world of contingency?

For example, when contemplating the truths of essence, it is of interest to note that one truth of essence is that the necessity of the truth of essence is different from the contingency of the truth of fact and that the difference between the two types of truth is itself a distinction that is necessarily true. (If the realm of contingent fact were to become transmogrified into a realm of essence, the distinction between the two realms would still be true in principle although it might be a distinction which was unknown and unknowable.)

Two questions present themselves. First of all, what implications follow concerning the existence of the world from the fact that a relationship of two realms of truth is known as an essence when one of the terms of the relationship is a term which refers to the contingent world? Secondly, is it of any significance that the idea of necessary essence itself, the could not be otherwiseness of essential truth might not be noticeable if it were not for the fact that it could be contrasted with the kind of truth that could be otherwise? The very possibility of noticing an epistemological distinction would depend in this case on the ontological existence of the contingent world.

These two very absorbing implications deserve to be further analyzed for the sake of drawing out possible consequences. With regard to the first implication, does it make a difference in the concept of a proof of the existence of the world if a necessary truth can be known concerning a relationship of distinction between levels of truth when one level of the distinction is obtained from a knowledge of the world? If the very knowledge of one kind of essence requires contingent existence, can it still be said that contingent existence falls completely outside of the realm of phenomenological interest? Further, if necessary truth can be known, can it be said that this offers no sign at all of the proof of the world's existence? If necessary truth cannot be known without a contrast to contingent truth, does this not in and of itself constitute a proof of the contingent world? Can it be that a proof of the existence of the contingent world has been constructed despite of, nay, because of, the insistence on maintaining the purity of the phenomenological standpoint? Is it the case that a knowledge of necesssary truth is at once a proof of the world?

From the fact that a truth can be known with necessity to be true wherein a part of the content which is known about is not a necessary truth of essence, does it follow that that part of the content which is not a necessary truth of essence must exist? It does seem to follow that in order to know that which is necessary to be necessary that the contingent would have to exist. But there is no necessity in knowing the necessary to be necessary.

Neither does it follow from the fact that a truth can be known with necessity to be true, that such a truth must exist. If essence and facticity are separable, then the knowledge of necessary existence does not entail its facticity. Nor does it follow that the knowledge of necessary existence entails that it does not possess facticity. Facticity and essence are incommensurable.

Could it not be said that the entire distinction between necessary truth and contingent truth is itself a necessary truth of a distinction which can be drawn in the realm of essence alone without reference to the realm of actual existence? Could not the concept of necessary truth be necessarily distinguished from the concept of contingent truth whether or not contingent truth possesses any valid ontological status? Could not contingent truth exist as a mere possibility? Or, could not contingent existence exist as an illusory possibility?

The world could be an illusion and not even known to be possibly an illusion, and still function as an appearance of contingency, or false contingency. The world could thus serve as an apparent other to necessity so as to permit the drawing of a distinction which is known as necessarily true without it having to be the case that the world would truly have to exist contingently. The contingent status of the world could be an illusion. Consciousness itself may be masquerading as contingency. The false appearance of a contingent world would suffice as a condition for the possibility of drawing a distinction between necessity and contingency.

Nevertheless, it appears that the necessity with which the distinction between necessity and contingency is known to exist would be dependent upon the *possibility* of the existence of the contingent world and therefore at the very least it would have been established that it is necessary that the world possess at least a possible existence. It appears that it must be the case that the contingent world exists either as a real or as a fraudulent contingency. But to say that it is necessary that it is possible that the world exists is the same as it is to say that the world may or may not exist; it does not follow that the world exists. Thus, once more, one finds oneself back where one started. The existence of the world cannot be known with necessity. Can it be said that a small difference, however, can be noted? Can it be said that the possible existence of the world falls entirely outside of the phenomenologist's province of knowledge? Must it now be said that at the very least it is true that in order that knowledge of the distinction between necessary existence and contingent existence is possible, that the possible existence of the world must be taken into consideration? While this may seem to be a very small difference, it is a difference and what is true, however tiny its truth value, must not be ignored.

Is it the case that the very concept of necessary existence itself could not be noticed without the possible existence of contingency? Is it possible that at last an argument has been devised which proves the possible existence of the contingent world? This would be ironic indeed coming at the hands of a philosopher who seemingly could care less whether or not the world existed or whether a proof could be devised which proved either that the world existed or did not exist. Nonetheless, it would remain true that the proof that the world either existed or did not exist would not be of great importance to the philosopher since such an existence would remain purely contingent and the realm of contingent truth possesses only limited interest for the philosopher.

However, given this above qualification, is it not true that a proof of the possibility of the existence of the contingent world has been devised in spite of the philosopher's reluctance and disinterest in offering such a proof or perhaps precisely because of the reluctance and disinterest that such a proof has at last become possible to devise? Unfortunately, while such a proof as the above may seem to be a proof, it is not itself foolproof. For, it may be possible, quite apart from the possibility of 'false contingency', to think of essence without contingency. The problem is, it is not knowable if it is possible to think of essence without contingent existence because the world in which the philosopher thinks already includes contingent existence. Thus, it is not knowable if necessity can be cognized on its own without contingent existence because the experimental thinking conditions are unavailable. That it appears that 'cannot be otherwiseness' cannot be thought without the possibility of 'otherwiseness' is only true in the world in which one finds oneself existing; it is not known or knowable if 'cannot be otherwiseness' would be capable of being known if the possibility of 'otherwiseness' were not known since this data base is unavailable to the philosopher. Therefore, while it may be true that necessity cannot be thought without the possibility of contingency, it is a truth that cannot be known.

If, however, one does not consider the possibility of a world existing with different experimental thought conditions than the world in which one already exists, and it is acknowledged that one can become aware of consciousness as that which must exist in order for one to become aware of consciousness and it is agreed that one is aware of consciousness, then one is aware of that which must exist if awareness is to be possible at all. If awareness of consciousness is possible as an awareness of certain truth, how can one be aware of the status of certainty, unless it is possible to contrast this certainty with that which is not certain? In order to become aware of the truth of consciousness, one must be aware of the possibility of the existence of something other than consciousness.

It seems that the possible existence of something other than consciousness may be fulfilled in three different ways: the possibility of the existence of that which possesses awareness, but no self-awareness; the possibility of existence which is not conscious at all, or brute facticity; the possibility of nothing at all existing outside of consciousness. Thus, to become aware of consciousness as a necessary awareness requires that one is aware of conscious minds, a physical world, the possibility of nothingness, some combination of or all of the preceding. The proof that the consciousness that is possessed is a necessary awareness requires that one is aware of other minds, a contingent world, and/or nothingness. Can it be that a proof has been devised which proves the existence of other minds? Can it be that a proof has been devised which is a proof of the existence of nothingness? Can it be that this proof may also be a proof that the world must either exist or not exist? Is this not ironic coming at the hands of a philosopher whose premises are that the phenomenologist possesses no concern in whether other minds or the world exists or do not exist?

It is difficult to say that anything at all has been proved. One must return time and time again to the safe haven of consciousness before one dares to venture out again. Most co-temporary philosophers shy away from this starting point or even find it to be an embarrassment, but this is a starting point that cannot be avoided and is avoided only with the consequence of the subsequent impoverishment of all further thought.

Unlike the more ancient and more mature Indian philosophy, in the West, with the major exception of Descartes, the starting point of the necessity of consciousness has been vastly under-utilised. By not having paid more attention to this starting point, philosophy's progress in the West has been severely retarded. Indeed, the directions which Western philosophy has taken have been wrongheaded and since they have been so multifarious and have occupied so many fine minds, the amount of regress has been astonishingly wasteful. One is reminded of Aquinas' remark in *On Being and Essence*, that, 'A small mistake in the beginning is a big mistake in the end'. Here, a big mistake in the beginning is a giant mistake in the end.

Philosophy must be returned to its starting point. How and why can any rigorous philosopher, in good conscience, leave this starting point or neglect its widespread consequences? The question which is raised is, is the necessary truth of consciousness the only truth of which one may be certain?[1] Is Descartes correct in maintaining that all other essential truths may be the products of a malicious daemon? Whether we are waking or dreaming is not the essential question as the illusion of essence can occur in either realm. The entire distinction between the waking and the dream state may itself be an illusion - it could be that there is no such distinction at all - both waking and sleeping can be equally illusory states when compared against a third state, which in Indian Vedandic philosophy is known as deep sleep, or pure consciousness, or *Turiya*. With a third reference point, the entire waking-dream debate can be seen to be a pseudo-conflict. However, it is a useful one since it is a harbinger of a more serious question, which is, if both waking and sleeping are equally unreal, then what is reality?[2]

While the world and the truths of inner consciousness are both realities which are subject to question, what is unquestionable is that a reality exists which has the nature of consciousness. Can it be said that the features of reality can be read, although indistinctly, in the mirror provided by consciousness? If consciousness is the only certain reality, then it is arguable that the products of consciousness can be relied upon as clues to the nature of reality. But is it true that consciousness itself is the only certain reality? Is the starting point provided by Descartes a sufficient starting point? Or, is it possible that the starting point for philosophy can be expanded?

For Descartes, any necessary essence, with the exception of the primary essent of consciousness, can be a product of a deceiver. Insofar as one is conscious, however, one can be aware of a possible deception regarding necessary essences. Of course, while it is possible that an omnipresent and omnipotent deceiver exists, it is not necessary to assume that an omnipresent and omnipotent deceiver exists by the mere virtue of its possibility. But the possibility of such an existence must not simply be ignored or discounted.[3] As Newton once remarked, he could see as far as he could because he was standing on the shoulders of giants. One must summon up the courage to mount Descartes' shoulders. Even if such an omnipresent and omnipotent deceiver exists, what Descartes has established is that a consciousness exists which is not dubitable (this is not to be confused with any particular empirical consciousness of any existent human being).

It is the task of the philosophy of the present and of the future to expand the list of necessarily known truths beyond the single known truth of consciousness. To borrow Descartes' device of the deceiving daemon, it can always be alleged that all necessary truths are but fraudulent clues left behind by the deceiving daemon. But it must be said that if everything is an illusion, then one still has to understand that. If this understanding is also an illusion, then it is meaningless to assert that everything is an illusion. At the very least, it is meaningful only to assert that everything apart from the understanding that everything is an illusion is an illusion. If one such understanding is possible, then at least one thing is true: it is true that it can be understood that everything except one's understanding that everything is an illusion is an illusion. Can it then be said that there is only one truth that can be understood? Is it possible that there could be only one truth? If everything is an illusion, is it possible to know this?

Without calling in the hypothesis of the good Deity, it is not true that there is only one thing that one cannot be mistaken about (e.g., the existence of consciousness). In fact, it is impossible that only one thing can be known to be true. If one analyzes the claim that one can only understand that every seeming truth apart from one's understanding that everything is an illusion, is an illusion, on the contrary, one can generate an indefinite number of certainties: (i) it is true that one understands that consciousness exists; (ii) it is true that one understands that it is true that consciousness exists; (iii) it is true that one understands that it is false that there is only one thing that is understood as true; (iv) it is true that one understands that an all pervasive illusion is incoherent; (v) it is true that one understands that one stands in truth, for unless one stands in truth and unless a stance is a true stance, there can be no sense in which an illusion can be perceived as an illusion; (vi) it is true that there exist five truths plus one additional truth (the truth that there exist five truths); (vii) there thus exist six truths plus one, *ad infinitem*.

The examination of the hypothesis of the omnipresent and omnipotent deceiver is not without value.[4] The examination of the hypothesis of the omnipresent and omnipotent deceiver leads to a greater trust in the powers of the human mind and to the knowledge that an infinite number of certain truths can be generated. To face directly the hypothesis of the omnipresent and omnipotent deceiver brings great rewards. Such rewards do not accrue to the undue and cowardly rejection of, resistance to, and ignoring of the entire possibility of the omnipresent and omnipotent deceiver which is the unfortunate legacy of present-day philosophy with which any philosopher now embarking upon her or his work must with a sense of sadness and repugnance realize is the pitiful starting point of the philosophy of the present era.

Notes

1. For Peter Schouls, Descartes must begin with '... absolute trustworthiness of some things known immediately [*notitia*, 'awareness of first principles']...' and it is clear from the context that he means that it is the *cogito* that is the first item of certain knowledge. Schouls refers to the *cogito* as the foundation for all further construction of *scientia*. (Thus, the term 'first' for all practical purposes in this context becomes equivalent to 'only'.) *Cf.*, 'Arnauld and the Modern Mind (the Fourth Objections as Indicative of Both Arnauld's Openness to and His Distance from Descartes', in Elmar J. Kramer, *Interpreting Arnauld*, Toronto: University of Toronto Press, 1996, pp. 33-63. In a revealing passage quoted by William Reid in his doctoral dissertation, Descartes refers to his knowledge of his being a thinking being as a primitive act of knowledge (*primum quaedum notio*) derived from no syllogistic reasoning. *Cf.*, William Lewis Reid, III, *An Examination of Descartes' Evil Demon Argument*, University Microfilms, 1979, p. 313. *Cf.*, Haldane and Ross, *The Complete Works of Descartes*, II, 38; AT VII, $140^{12\text{-}19}$. In the same work, Reid relates Popkin's interesting historical account of the possible origin of Descartes' argument in the trial of a parish priest, Grandier, in Loudun, who was accused of infesting the local convent with devils. If Grandier were deluded, then the detection of his delusion through examining him would have been impossible. *Cf.*, *Op. cit.*, pp. 13-14 and Richard Popkin, *The History of Scepticism from Erasmus to Descartes*, New York: Harper and Row, 1964, pp. 182, 185. It is amazing to consider that such a powerful philosophical argument can have been extracted from an eccentric, empirical case in history.

2. While Plato does not raise the argument of a malignant daemon, he does say in *Theatetus* that there is no evidence to which one may point to prove whether we are awake or that all that passes through our minds is a dream (158B, C).
3. In his book on Descartes, Bernard Williams points to the argument of Margaret MacDonald that one could always be dreaming that one were satisfied that one had found the criterion of distinction between waking and dreaming. In his own efforts at a solution (which Plato had argued to be impossible), Williams argues that from a waking state one can explain dreaming, and that this is an important asymmetry, but this surely begs the question. Williams' argument seems to take the form that it is illegitimate to infer from the possibility that one may be mistaken about when one is awake that one is therefore always so mistaken. But the point of Plato and Descartes (prior to *Meditation* VI) is that since one may at any time be mistaken, it does not follow that one is always mistaken but what follows is that one may never be sure on which occasion one is mistaken and which not. This is the point that cannot be gainsaid. *Cf.*, Bernard Williams, *Descartes, The Project of Pure Enquiry*, Harmondsworth: Penguin Books, 1986, pp. 309-313. Williams himself appears to supply an even more intriguing point when he argues that we cannot rely upon the veridicality of a past dreaming state since even the thought that we may have been dreaming is suspect (*Ibid.*, pp. 57-8). Williams seems to be implying that the dream argument is in fact a provisional and less hyperbolical form of the malicious daemon argument.
4. In a remarkable sentence, Schouls comments, 'Counterbalancing a "supremely good" G-d with a "malicious demon of the utmost power and cunning ..." has the effect of freeing the thinker from both'. *Cf.*, Elmar J. Kramer, *Interpreting Arnauld*, Toronto: University of Toronto Press, 1996, p. 44. Most thinkers, however, including those of Descartes' time and those of the present time, do not wish to take Descartes' radical step. For example, consider this statement of Dicker's: 'Most philosophers today, including the present writer, would firmly reject Descartes' doctrine of the creation of the eternal truths. They would hold, as Leibniz held, that the truths of logic and pure mathematics are true in all the possible worlds that G-d could have created, so that not even an omnipotent G-d could dictate or alter them. They would also agree with the view, put forward by Aquinas, that omnipotence does not require the power to do logically impossible things, but only the power to do whatever is logically possible.' *Cf.*, George Dicker, *Descartes, An Analytical and Historical Introduction*, Toronto: University of Toronto Press, 1996, p. 136. It is surprising that there is so much resistance to taking Descartes' radical step. And, it is interesting to reflect that Descartes, heralded as the arch rationalist, does not think that G-d should be limited by the laws of logic. Dostoevsky grants more freedom to human beings than Leibniz, Dicker and company would ever grant to G-d. One need only recall Dostoevsky's passage in *Notes From Underground* in which he states that, 'I admit that twice two

makes four is an excellent thing, but if we are to give everything its due, twice two makes five is sometimes a very charming thing too'. Descartes makes bold to grant G-d just such power. In a letter to Arnauld, Descartes states that, 'I would not dare to say that G-d cannot make ... [it] that one and one should not be three'. Descartes becomes a veritable apostle of post-Modernism! Indeed, in other letters he states that, 'G-d cannot have been determined to make it true that contradictories cannot be true together, and therefore ... He could have done the opposite' (to Mesland, 2 May 1644). And, 'Just as [G-d] was free not to create the world, so He was no less free to make it untrue that all the lines drawn from the center of a circle to its circumference are equal' (to Mersenne, 17 May 1638). Furthermore, 'The mathematical truths which you call eternal have been laid down by G-d and depend on Him entirely no less than the rest of His creatures. Indeed to say that these truths are independent of G-d is to talk of Him as if he were Jupiter or Saturn and to subject Him to the Styx and the Fates.' (to Mersenne, 15 April 1630), *Cf.*, Dicker, p. 135. As is argued above, one does not take Descartes' step at the peril of the weakening and the impoverishing of philosophy. There is at least one case in which a solution presents itself in such a way that one may rescue G-d's omnipotence and not violate the laws of logic. This is the case of the problem of whether G-d can make a boulder so heavy that it is beyond the Divine power to lift it. In this case, it may be argued that such a problem is a pseudo-problem since the most current power would always define the Divine essence. Thus, the current power of lifting is the defining characteristic of G-d and not His previous power of making. That He cannot make a boulder so heavy that He cannot lift it is not a limitation of Divine power; it is a proof of His powers of Self-surpassing. The Divine always increases in power and the problem arises from the limitation of the concept of omnipotence when it is circumscribed by the notion of the necessity of complete temporal actualization, and does not reflect at all on the limitation of G-d's omnipotence. In addition, to all such questions as 'Could G-d have commanded a creature to hate him?' (Burman's question), it must be asked, not only could He, but would He? To omit the 'would' is to make G-d into a purely logical being without ethical characteristics. One could say that G-d could have so commanded but that He would not have done so. This is different than the argument of Descartes which seems to veer away from granting G-d total freedom when he answers Burman that 'God could not now do this, but we simply do not know what he could have done'. In Garntsev's construction of Descartes, G-d is thus constricted in His countervening the laws of common life that he has already established. *Cf.*, Mikhail Garntsev, 'The Problem of Absolute Freedom in Descartes,' *Russian Studies in Philosophy*, vol. 37, no. 2, Fall, 1998, pp. 7-17.

Allegro deciso
Rapid and flowing

10 The Intuition Taboo

While two sides of the *cogitatum* may be logically abstracted into *noesis* and *noema*, such a separation does not exist at the moment of knowledge. By starting with the desire to know, Aristotle is surprisingly modern. In fact, by beginning with the desire for, rather than the fact of awareness, he resembles Freud even more than Descartes, in his anticipation of modernity.

With his famous first sentence of his *Metaphysics*, 'All men by nature desire to know', Aristotle appears to be beginning with an assumption or a general truth from which he will deduce conclusions.[1] Of course, it could also be inferred that Aristotle arrived at this conclusion from a process of simple observation. On the other hand, no set of observations would be adequate to form a conclusion that applied to all human beings. How does one legitimately arrive at such a conclusion? To conclude Part I, it is only necessary to offer a phenomenological interpolation of Aristotle's famous statement that all men by nature desire to know. It could be argued that here Aristotle does not begin with an assumption from which he deduces consequences. It could be argued that he begins with a reflective truth ('All men by nature desire to know') that is immediately apprehended, which is the reflected truth of what is known, or the *noemata*. One begins with philosophical truth; it is there at the very outset. It is, after all, his first sentence. He does not begin with an explanation of how, after observing many men, he has arrived at this inference as an empirical generalization. He simply begins with this sentence as if it were an obvious truth. It is entirely possible that such a truth appeared obvious to Aristotle because it was, for Aristotle, a phenomenological truth. It was a truth that reflected his own introspection of his own consciousness. At least, since Aristotle does not relate how he has arrived at this famous conclusion, the path of conjecture is open. In any case, regardless of how Aristotle has arrived at this conclusion, Aristotle is more philosophically courageous than Kant. Kant begins with the truths of other disciplines; Aristotle begins with the truths of philosophy.

It is with the existence of these kinds of truths that the philosopher is primarily concerned. How these truths are known to be true is another question and requires a secondary act of reflection. What is of importance is the first discovery of the truth. Philosophical truth is objective and not subjective in that the truths which are known are known as objective truths whereas the venue in which they are known is the immediacy of subjective introspection.

It may be said that it is with the fact of subjective consciousness that the phenomenologist begins, but this is a methodological truth (a part of prediscovery, or, the hermeneutic base). While it is perfectly true to say that subjective universality is what is meant and what can only be meant by objectivity, it is equally true that such subjective universality is only possible by virtue of the fact that philosophical truths are discovered and are by no means invented. The subjective condition of truth discovery is not the condition of the verification of the truths of internal consciousness; that is due to the self-verifying nature of the objective traits of the truths in question, and the fact that these are accessible through consciousness is only incidental.

As a result, the prevalent notion that Modern philosophy begins with the subjective, as in the case of Descartes, is misleading. Whether one refers to Descartes or to Aristotle, both begin with the objective truths of consciousness, and it is the particularities of the truth contents or the *noematic* side and the implications by analogy that may be drawn from these truths that is the interest of philosophy to explore. Philosophy is only and always involved with objective truth, and not subjective truth, whether it be in the philosophy of Aristotle or Descartes.

What is perhaps amusing is that whereas Greek philosophy is construed as beginning with the objective, from the point of Aristotle's actual starting point in *Metaphysics*, he begins (from the order of exposition) with a statement concerning human beings and their desires for knowledge. Aristotle begins with the desires of the subject knower. For the phenomenological reader of Aristotle, the means by which Aristotle's statement is known to be true is by each reader's internal reflection which she or he discovers within her or his own consciousness.[2] While the content of Aristotle's statement is indeed objective (as the *noema* is the truth content of a philosophical proposition), it has not been understood to be objective for the right reasons. In the case of Descartes, while he is considered for his beginning with subjective consciousness to be the father of Modern philosophy, his starting point is in fact no different than that of Aristotle's, and it is the *noemata* he discovers that remain of greatest interest, and it is by means of the discovery of these that his philosophy also achieves greatness. In his own terminology he is concerned with what are objects of consciousness (thereby, objective contents in the meaning of objective which is employed here), or what has also been referred to here as *noemata*. All philosophy worthy of the name has always begun with the objective, and all philosophical truths are objective truths, including the truths concerning one's *noetic* powers, whether or not they happen to exist within the domain of, or even form the very basis of subjective consciousness. Thus, 'All men by nature desire to know', is a *noematic* truth from the standpoint of philosophical classification, although what it informs upon is the background and the extent of a *noesis*. Intuition or in the terms of the present author, transcendental introspection is the means by which the above statement of Aristotle's is known at least by the reader of Aristotle to be true. What is discovered *in* though not *by* transcendental introspection are *noematic* truths of philosophy and thus objective truths, even though the term 'intuition' refers in the popular sense to a subjective, and usually thereby erroneously considered to be a variable, undependable, and unverifiable faculty.

To conclude, one must not understand from the foregoing either that everyone must possess some power of intuition, or, that it is owing to the power of intuition that the statement 'All men by nature desire to know' may be verified to be true. This is exactly the wrong conclusion to reach, and reflects how philosophy is all too often so badly misunderstood. Aristotle's statement is known to be true because of the property of the statement as being true. It is from the side of the *noema* that truth is discovered in philosophy. If it were otherwise, then intuition would in fact be the culprit monger that it so often is taken to be. It is the presence of objective truths in philosophy that make philosophy what it is. If one bows to the need to coddle the popular mind by attributing truth to the power of intuition, then the *intrinsic content* and the *importance* of the philosophical truths that are *discovered* and that they are and *must be discovered* is lost. This is, perhaps, why philosophy does not command the respect that is its due. It is from the presence of the philosophical truths that remain to be discovered that the attraction of philosophy should take its rise. A philosophical audience may enjoy the truth-value of the truths that have already been discovered as in this chapter, but this is merely an *apéritif.* The real work of philosophy lies ahead and in this sense, the real achievements of philosophy belong to the philosophy of the future. This work serves only as a prelude to the philosophy of the future, and is dedicated as a propadeutic to all future philosophy, so that philosophy can be restored to its position of respect among the disciplines of mankind.

Notes

1. One is reminded of that well-known passage in the *Phaedo* in which Socrates exclaims his delight in hearing for the first time that Anaxagoras has said, as someone who is reading from a book relates to him, that everything is caused by mind. Socrates' knowledge is occasioned by pleasure; in fact, the basis of his consideration that Anaxagoras has arrived at a great truth is the pleasure with which his mind takes in his first hearing the idea. It is entirely possible that Aristotle arrived at his own connection between desire and truth from hearing his master Plato's discourses of Socrates.
2. A commentator once criticized Aristotle for delivering only 'dazzling glimpses of the obvious'. This criticism could be extended to all philosophical discoveries such as the Cartesian discovery of the *cogito*. The problem with this criticism is that what is obvious becomes obvious only after it is seen. The immense problem is that human beings generally fail to see what is obvious. The proper meaning of the obvious is that what is obvious is what should be seen but is normally not noticed. To be given dazzling glimpses of what one normally fails to notice is a gratuitous gift. In Hegel's famous phrase, '*Das Bekannte uberhaupt is darum, weil es bekannt ist, nicht erkannt*' (What is familiar is not necessarily known). Or, to adapt this phrase, what is the most familiar may be the least well known.

PART II

The Nature of Ethics and the Bio-psychological Deduction of the Emotions

Moderato espressivo (il tocco dolce, ma pieno)
Quietly movemented and expressive (with a soft yet full touch)

1 Toward a New Ethical Foundation: Bio-evolutionary Social Theory

The Original Goodness of Human Nature

I. Methodological Considerations

A. The Problem of the Type of Existence of the Subject Matter

The first question, which might occur to the student of human nature, is whether the inquiry is to be a descriptive or a prescriptive one. One may commence by questioning the soundness of the descriptive/prescriptive distinction. Prescriptions must relate to the basic potential to fulfill them. Chinese philosophy has been both sharper and earlier on this point than Western philosophy. For example, Mencius, unlike Kant, considered that morality must be based on man's nature in order to be practical. In his reply to Kao Tzu, Mencius says:

> Sir, can you follow the nature of the willow tree, and make the cups and baskets? Or must you violate its nature to make the cups and baskets?[1]

Aristotle placed great value in the explanatory power of the relationship between potency and actuality. But despite his use of these concepts, he apparently saw no difficulties in holding that man's nature was evil while his ethical potential was realized in performing good acts.

The position taken up here is that the inquiry is both a descriptive and a prescriptive one. The descriptive element is the existence of a nature; the prescriptive element is the aspect of the nature that requires guidance and direction. The prescriptive element must relate to the descriptive element such that what is or can be actualized is congruent with the most basic potential of the nature.

B. The Problem of Access to the Subject Matter

Another question, which must arise for any student of human nature, is how one can lay claim to having any access to a pure datum. This is the problem of culture. Every human nature taken as a datum is already influenced by some culture. Every investigator is influenced by some culture even if she or he could come upon any pure experimental datum. The investigator is influenced by the data and the data are influenced in turn by the investigator: a two-way-Heisenberg effect.

First of all, it is important to refuse to accept to some extent, the absolute existence of the problem. The assumption behind the posing of the problem is that real human nature would exist apart from or aside of any cultural context. Contrariwise, it may be arguable that real human nature, at least at some point in history, must always appear from within some cultural context. An experimental Skinner box may distort human nature rather than afford us a vision of its pure essence. In other words, there is no human zoo; or, culture is the human zoo.

Secondly, it does not follow that if nature appears within a cultural context that its essence is derived from that cultural context. It is arguable that, to some extent, there is a basic human nature that cuts across all cultural barriers and is discoverable by the investigator.

Lastly, it is part of human nature to become part of human culture. It is part of human nature to influence itself in a certain direction. The descriptive and prescriptive elements are not totally unrelated. The investigator cum philosopher has a responsibility to select and emphasize those features of human nature that most closely answer to its essence. The prescriptive element must be based on the descriptive and can also modify the descriptive so that the descriptive becomes the prescriptive and vice-versa.

II. Substantive Considerations

The basic inspiration for the outlook on human nature that is adopted here is a core insight of Chinese philosophy that human nature is inherently good.[2] This basic inspiration may be reaffirmed from three standpoints: metaphysical, biological and social. In so far as the intent here is to concentrate on human nature rather than on nature in general, biological and social standpoints are the focus of concern, which is why this theory of ethics may be labeled the bio-evolutionary social theory. While it has been argued in this present work that philosophy should not take its rise from theoretical foundations, in the case of claims that contain a prescriptive element, theory cannot be totally avoided. One attempts to find a prescription which closely corresponds with the most positive direction of the descriptive essence.

If man's nature is inherently good, then her or his biological nature is inherently good. If the master key to man's biological nature could be isolated, it should provide one with a significant clue to her or his ethical nature. The master key to man's biological nature is that it is directed towards the preservation of the human species. All human motivations and actions must find their ultimate grounding here. All social evolution can be seen as an extension of the most basic drive of species preservation inherent in the human being.

A. Metaphysical

From the standpoint of metaphysics, every nature must be itself. Its essence cannot be not to be itself or it would not even be in the first place. If it could be, it would not last for very long. If its essence is to be itself (whatever it is), then it is also its essence to maintain itself in being.[3] A nature cannot be (certainly not for very long) if its basic *telôs* is not to be itself or to destroy itself. From the standpoint of a metaphysical concept of a nature, human nature cannot be inherently constituted to destroy itself. The basic principle of human nature cannot be other than self-preservative.

B. Biological

The most constitutive *telôs* of any biological species is the principle of self-preservation. Human nature cannot be designed for intra-species conflict because such a design would be at the very least inefficient and at the most ultimately self-destructive. It makes no sense, from a biological standpoint, that a species should be so constituted as to be inherently self-destructive. Survival, rather than destruction, is the *telôs* of a biological species. Survival, in terms of the evolutionary theory presented here, must be understood as survival not of the "fittest", but of the species. If evolution were a struggle simply for survival, why would it not have ruthlessly eliminated all altruists, who seem to increase another's prospects for survival at the cost of their own? [4]

There is no basic instinct for aggression or death wish in human nature. Aggression, as is argued below, is always in service of life. The

Western view of two opposing forces, whether Manichaean good versus evil or Freudian *Eros* versus *Thanatos* is not well grounded from the biological standpoint.

C. Social

Human nature cannot be conceived after the model of a human individual. No human being can even come into existence as an individual. Minimally speaking, it requires a prior society of two in order that one human individual exist in the first place. Human nature is already ontologically social.

If one combines metaphysics with biology, again one reaches the conclusion that human nature cannot be designed for social conflict. If human nature must be social in order to be at all, it cannot be so constituted as to be inherently antisocial. Human nature is designed for social harmony. This is again another core insight of Chinese philosophy, which derives from the fundamental insight that human nature is originally good. Original goodness is required as a prerequisite for social survival. This is a kind of biological equivalent for ethics. This is not to say that an ethical theory must be consequentialist or teleological to be valid. It is also not to say that ethics may not possess its own intrinsic value. It is only to say that ethics can be grounded on bio-evolutionary social theory.

D. Bio-psychology

The core drive of the human species (or any species for that matter) is for self-preservation, the philosophical expression of which is life affirmation. Self-preservation or survival must of course include reproduction, as the species would not survive if it did not perpetuate itself in addition to conserving its present generation. Biological self-preservation cannot function on its own: it requires the instrumentality of psychological drives for its implementation. The following description of psychological drives is meant in a very general sense and is naturally much condensed. For purposes of brevity certain physiological and psychological drives have been collapsed. The terms utilized are thus to be understood primarily in a general philosophical sense rather than in the more limited physiological or psychological sense even when they are specifically so designated. Life affirmation does not mean, of course, physical survival at any cost. Life affirmation takes for granted that certain values must also be preserved.

The Bio-psychological Deduction of the Emotions: The Theoretical Foundations for the Grounding of a New Ethics

1. Self-love, self-esteem, self-care, self-respect, or the love of the species for itself is the most basic psychological drive and is the psychological expression of the core biological drive of the human species for self-preservation. Self-preservation manifests itself in two forms: (i) conservation; (ii) perpetuation.

1. i. The physiological expression of the conservation of the species is the appetite of hunger.

1. ii. The physiological expression of the perpetuation of the species is the appetite of sex. The psychological correlate of sexual appetite is sexual drive.

1. iii. When the physiological expression of the conservation or perpetuation of the species is satisfied, the result is happiness. Happiness can also be experienced when one is the recipient of another's desire for the perpetuation of the species such as when one experiences maternal love or, in Japanese cultures, *amae*. Higher orders of happiness may require the satisfaction of higher desires such as aesthetic desires, religious desires, intellectual desires, desires for meaning in life, desires for friendship, and/or desires for small group, community or national affiliation. At a point in social evolution, human happiness may be identified with the capacity for the satisfaction of higher desires such that the possibility of the non-satisfaction of those desires may require the sacrifice of human life. Hence, the case of the noble self-sacrifice of all the Jews at Masada. Survival of the species may also take the form of attempts to preserve higher forms of culture at the cost of individual survival.

2. When the well being of the object of self-love is threatened, the natural response of the human organism is fear and/or the desire to protect the object of self-love. (Laterally, the natural response of the human organism is hope, the contrary to fear. For the present purposes of the bio-deduction, the path of fear is followed and not of hope. This is not to say that hope is less important than fear but to trace its derivations would interfere with the present discussion). Fear is derivative from the threat of diminution, hurt, loss or the diminution, hurt or loss (real or imagined) of the object of self-love. Fear manifests itself in one of three modes: fighting, freezing or fleeing. Each mode is an attempt to remove the threat to survival.

2. i. Fighting: the response to remove the threatening object by attack. Attack is always defense. Every aggressive act is a fear response or a protective response.

2. ii. Freezing: the response to remove the threatened self via mental avoidance (the object of threat is removed from the mind by panic) accompanied by the lack of physical movement (paralysis) or pretense accompanied by the lack of physical movement (as by playing possum). The psychological form of freezing is manifested as denial.

2. iii. Fleeing: the response to remove the self from the threat by physical withdrawal via physical or mental movement to escape or hide.

2. iv. One example of a mental response of hiding is the narcissistic personality disorder in which the narcissist hides from his or her feelings of low self-esteem by a compensatory self-inflation. In this case, the narcissist flees from a devalued sense of self to lose himself or herself in an overvalued sense of self and unwarranted conceit. In extreme cases, the narcissistic self-inflation can lead to grandiosity and even delusions of grandeur. Grandiosity is a masque for depression. The narcissist can also flee to hypochondriacal anxiety in which the flight from self is a flight to an obsession with psychosomatic symptoms. Paradoxically, the narcissist can display an absence of affect, which is in reality the result of the containment of intense feeling. Alternately, or simultaneously, the narcissist can transfer the feeling onto the therapist through psychic bleeding (as is common with the borderline personality disorder).[5] The narcissist can move from flight to fight by taking offense at the slightest provocation.

2. v. All psychopathology can be interpreted as a flight from the possibility of the rejection of the self or from intense involvements, which betoken either a swallowing of the self by the other or an abandonment of the self by the other. For example, schizophrenia can be characterized as an overt withdrawal from threatening objects.

3. When the organism responds to remove the threatening object, fear turns into anger or hatred. When the organism fails to remove the threatening object, fear turns into anger against oneself for the failure to take action. Such anger against oneself manifests itself in guilt or shame depending upon the cultural background for one's early development. (Western cultures are predominantly guilt cultures; Eastern cultures such as Chinese and Japanese cultures are predominantly shame cultures.) If there is no expiation for the feelings of guilt or shame, guilt or shame turns into depression. If depression is allowed to continue for a long time, it may turn into self-hatred, resignation, despondency, hopelessness, some combination of, or all of the above.

3. i. Anger is the response of the organism to a short-term threat. Frustration is anger in an incipient stage.

3. ii. Hatred is the fear response of the organism to a long-term threat.[6] Since hatred is at least three steps removed from love, it is

obvious that it is not love's opposite or as some would have it, part of the meaning of love. Love's opposite, or to speak more precisely, love's privative, is not hatred, but fear.

3. iii. Sadness is the response of the organism to the actual or the imagined hurt inflicted upon or the actual loss or diminution of the object of self-love, or, when the object of self-love is another, the response to the actual or the imagined hurt or destruction of the object of self-love.

3. iv. Hurt is the response of the organism to the actual or imagined hurt inflicted upon the object of self-love, or, when the object of self-love is another, the response to the actual or the imagined hurt or destruction of the object of self-love.

4. When fear responds by removing the threatening object whether directly through attack or indirectly by refusing aid, fear can turn into guilt or shame. Every case of guilt is actually a special case of shame in which the fear in this case is of the opinion of the self regarded as other.

4.i. When fear responds by removing the threatened self, fear turns into shame and resentment. Resentment sooner or later must turn into anger or hatred. When resentment turns into anger or hatred it becomes vengeance. Self-punishment is vengeance directed against oneself.

5. Greed is hunger mixed with fear that the organism is not receiving enough for its survival, present or future.

6. Envy is a modality, which exists in the transition state between fear and hatred. With respect to fear, envy is the fear that one will be outshone by the other. With respect to hatred, envy is a special case of hatred in which the object of desire is possessed by another but the greed threshold has not yet been passed.

7. Jealousy is a special case of hatred in which the object of desire is possessed by another, but the greed threshold has been passed. Jealousy is divided into two forms:

7. i. Simple jealousy is the state in which one does not want what the other possesses, but one does not want them to possess it either.

7. ii. Covetousness is a case of jealousy in which one also wants to possess what the other possesses. In the Biblical story, King David coveted Bathsheeba.

It would be of enormous interest to continue with such derivations. For example, the deductions could be expanded into the realm of specific unethical acts. Lying or promise-breaking can both be seen as special cases of attacking (the attempt to perpetrate injury by damaging or discrediting the threatening object), or freezing or fleeing (to avoid the threat of violence). All cases of lying, in short, would be seen as arising out of either fear, the desire to protect, special cases of hurt transformed into anger, hatred, or envy (as in the case of spreading malicious and/or slanderous gossip). The entire typology of character disorders can be displayed as in 2. iv. However, the list of deductions must be cut short in order to carry through with the present argument.

When anger or hatred erupt into the act of removing the threatening object, the result may be injury to the threatening object. The ethical evaluation of such an injury done to another is that such as action is 'evil'. (Even when the injury is only intended its 'evil' or 'wrongness' is also assayed in terms of the injury that would have been caused to the other.)

What is most important to note is that all emotions are derivative from the master emotion of self-love. Anger or hatred, in particular, are derivative from the deprivation or threatened deprivation of self-love (by the threat to the proper object of self-love). Acts, which are labeled as evil, arise from emotions, which only arise when the existence of human nature is threatened. *The conclusion, which can be drawn from this, is that evil or the instrumentality of evil (e.g., hatred, anger) is not present in original or undisturbed human nature.*

Thus, Hsün-Tzu is mistaken when he asserts that '... man is born with envy and hate'. In the above analysis, hatred or envy are not part of original human nature but only arise when the survival of that nature is under threat. Hsün-Tzu's full statement is as follows:

> The nature of man is evil; his goodness is acquired. His nature being what it is, man is born, first, with a desire for gain. If this desire is followed, strife will result and courtesy will disappear. Second, man is born with envy and hate.[7]

Hsün-Tzu is not quite correct when he states that man is born with a desire for gain. What he has identified as a desire for gain is either a legitimate expression of self-love or hunger, or greed, which is self-love affected by fear (the fear that one will not get or possess enough). If one attempts to satisfy hunger, the natural result is not strife. Under social conditions of scarcity, the social result may but need not necessarily be strife. But if strife occurs, this is not as the result of inherent malice; the occurrence of strife may be a purely social problem, which requires a social solution, such as improved production, distribution and an ethical economics.

Hsün-Tzu is correct in tracing the origin of evil (at least in part), to certain emotional states; it is only that he has not recognized that these states are not inherent to human nature. If Hsün-Tzu were correct and man were born with envy and hate, there would be more of a chance that the rest of his argument could be correct. As it is, what he has identified as part of human nature is a second derivative (2, 3.i.). It would be useful to revisit an example from the other side, the famous example of Mencius of the child about to fall into a well:

> Why I say all men have a sense of commiseration is this: here is a man who suddenly notices a child about to fall into a well. Invariably he will feel a sense of alarm and compassion. And this is not for the purpose of gaining the favor of the child's parents, or seeking the approbation of his neighbors and friends or for fear of blame should he fail to rescue it.[8]

While Mencius takes this as an argument for an inherent sense of compassion, this sense of compassion can be taken as a special instance of the inherent core drive of species preservation. This is not to suggest that this is how it is intended to be taken by Mencius. But if human nature is inherently good, and one construes this nature to be biological nature, then not only should ethics and biology not be incompatible, they should be positively correlated.[9] In fact, if one wishes to make an addition to Mencius' idea, one can consider both the phenomenon of the spontaneous benevolent action towards a fellow human being in danger and the feeling of happiness or contentment that follows upon the spontaneous exercise of the fundamental inclination to help one's fellow human being. That one may engage in a spontaneous act to help a fellow human being may be taken as an indication that such an act is in keeping with one's inherent nature since no deliberation need or in fact can precede such an act. That one feels a positive sense of contentment upon the performance of such an act can be taken as a sign that such an act was in keeping with the fulfillment of one's nature in the same sense that a feeling of contentment after eating may be taken as a sign that eating was in keeping with one's nature. One may even take the spontaneous feeling that follows upon an action of gratuitous benevolence as a sign that the action itself must have been the result of an inherent disposition to act and that such an action was followed by a positive feeling as a sign that such an inclination was one which nature wished to reinforce in human beings by rewarding human beings with pleasure at the performance of such an act.

That someone might thereby object that one was performing ethical acts for the sake of the pleasure that they brought, and therefore the acts were not performed for their own sake (or for the sake of helping others), but were performed for the selfish pleasure that they brought, and were therefore selfish, and not selfless acts, would be to miss the point that the acts, when performed, were performed *spontaneously* and without thought to the possible gain of happiness that was thereby to be derived (which the performer of the act would not even realize would be a side effect). And even if it were to be argued that one could be performing such acts for the sake of feeling good (or to relieve oneself of the guilt of not performing such acts in general), the point remains that one has still performed an act to help another whether or not one has derived pleasure (or pain) thereby.[10] If it were in one's nature to be benevolent, then it is consistent that one should feel content when one fulfills one's nature; it would be painful to commit benevolent acts only if such acts were contrary to one's fundamental nature.

It has been objected that the example of Mencius is flawed, because it is merely a reflection of empirical culture rather than nature. How can one know that the reaction is or is not culturally acquired? Consider the ghoulish case of Nazis seeing Jewish children falling into wells, or what is more likely, actively throwing Jewish children into wells, who would 'naturally' react entirely differently.

In defense of Mencius, it may be said that the reaction to save the child, whether or not culturally taught, is congruent with original human nature. If, in some cultures, the reaction of culture is identical to the reaction of nature, it may be said that that culture is congruent with original human nature. If, in some cultures, the reaction is different from that of nature, it can be said that this is a case where a culture is incongruent with human nature. The example of Mencius may or may not be supportive of how all people will empirically feel, but most of all it is supportive of how human nature's first reaction from the biological standpoint is for the preservation of its own species. Special cases such as the practice of systematic extermination of Jewry by Nazi Germany can be argued to be nonetheless guided by some notion, however horrifyingly distorted in the case of the Nazis' paranoid and morally bankrupt delusion, of the imagined preservation of their narrow and limited notion of the 'species' (which was not the human species in general as was obvious). The case of the Nazi mass genocide of Jewry is a clear case where the hatred of a group acts to extirpate that threat (real or imagined - in this case clearly imagined) to its survival.

What is important to realize is that neither hatred nor fear are ontological realities in their own right; both hatred and fear are parasitic upon love. When survival is assured, fear disappears. Fear only arises when there is a threat to the object of love. Fear has no existence in its own right. Its status is derivative and ontologically dependent. Fear cannot exist without the existence of an object of love. Fear and its *modus operandi* of anger, hostility, aggressiveness, envy, drive for power over the other, desire for revenge, self-punishment (revenge against the self), hatred and its special forms of racism, sexism, ethnocentrism, religiocentrism, are one and all dependent upon the existence of love for their existence.

Love, on the other hand, does not require fear for its existence. It is perfectly capable of existing without any dependence upon fear. Love is the primary reality because the species must care for itself, if it is to exist at all. An interesting example of this is the advice offered on airlines that in emergency conditions, when it is necessary for the oxygen masks to be utilized, that one should always take the oxygen mask for oneself first and then take an oxygen mask for an accompanying child. The point of this advice is that one cannot assist another unless one has first taken care of oneself. Taking care of oneself first is not selfish; the act of taking the oxygen mask for oneself first is an illustration of how self-love is a necessary condition for species preservation. Sometimes self-sacrifice is indicated as an instance of self-love. For example, females and children are to be saved first in cases of disaster, as with the case of a sinking vessel. In these cases, a male person may choose self-sacrifice in the interest of the preservation of the species.

Care for the self and the species is the primary or basic emotion and is the condition for the existence of all other emotions. Hatred can only exist so long as there is an object of love, which is threatened. When there is no frustration or impediment to love, hatred and its multifarious forms have no existence. Its only existence is as a derivative existence to love, when love is denied. It has no positive existence in its own right. Love, on the other hand, does not require the existence of hate for its own reality. It is perfectly capable of existing in the absence of hatred. Hatred and fear cannot exist except in the absence of love, since hatred and fear are simply forms of species self-love or individual self-love denied.[11]

Social Evolution

In closing, it may not be amiss to say a word about the 'progress' in social evolution. In bio-evolutionary social theory, however sophisticated mankind becomes, social changes must in turn serve primary biological drives.[12] At this moment or at this stage in social evolution, man runs the serious risk of species destruction in the short run with his supreme weapon of fear, the nuclear warhead, or in the long run with the by-products of greed derived from fear: environmental pollution and its effects on the planet. The lesson of social evolution is that fear, greed and conflict have led humanity to the brink of destruction in a game of species survival brinkmanship.

If man continues to operate out of derivative emotional responses of fear or hatred, the prognosis for species preservation is not good. What is important is the capacity to trust in human nature, and this is not possible if one subscribes to the belief system that human nature is malicious. It is interesting to speculate whether it is simply due to pure coincidence that the nations which first developed the supreme weapon are also those nations that share this belief system.

It has been the intention of this outline of the theoretical foundations of a new ethics to provide some bio-evolutionary social deductions; in particular the bio-psychological derivation of emotions, to suggest that original human nature is devoid of malice.[13] It is human nature to be socially harmonious. It is more important today, than ever before, to take notice of this great insight that may be taken as the proper heritage of Chinese philosophy, that man is not antisocial by nature.[14] For social harmony is a prerequisite to species survival.

Notes

1. *Mencius*, VI A:1. Wing-tsit Chan's translation in Wm. Theodore de Bary, (ed.), *Sources of Chinese Tradition*, New York: Columbia University Press, 1960, p. 102.
2. *Cf.*, Robert E. Allinson, 'An Overview of the Chinese Mind', in Robert E. Allinson, (ed.), *Understanding the Chinese Mind: The Philosophical Roots*, New York: Oxford University Press, 2000, Tenth Impression, pp.1-25. *Cf.*, Chung-ying Cheng, 'Chinese Philosophy: a characterization', in Arne Naess and Alastair Hannay, (eds.), *Invitation to Chinese Philosophy*, Oslo: Scandinavian University Books, 1972, p. 148 *et passim*. While Professor Cheng has specific reference to the Confucian tradition in his statements regarding the inherent goodness of human nature, he also seems to refer to the concept of the goodness of human nature when characterizing Chinese philosophy as a whole. At the very least, it can be asserted that this emphasis on the goodness of human nature is a unique feature of the mainstream of Chinese philosophy. The classifier 'inherent' is preferable to 'innate'. While 'innate' could conceivably include disposition in a broad sense, 'inherent' is more unambiguous in its meaning. 'Inherent' can be understood to stand for that which inheres as a fixed quality, without which the organism would not and could not be what it is. Thus, a disposition to respond aggressively might be innate

without being inherent (even if one take aggression as non-malign). *Cf.*, *Webster's* 2nd Ed.: def. 2 for Inherent: 'Involved in the constitution or essential character of anything ... inalienable ...' (thus it cannot be taken away). In addition, 'inherent' is intrinsically more suitable as a philosophical term since 'innate' might also carry certain physiological associations. The view of inherent goodness also goes further than such viewpoints as those advocated by Montagu, Fromm, *et. al.* Montagu thinks that aggression is a matter of individual variation whereas Fromm allows for non-malignant aggression. *Cf.*, Ashley Montagu, *The Nature of Human Aggression*, New York: Oxford University Press, 1976, pp. 9-11 *et passim* and Erich Fromm, *The Anatomy of Human Destructiveness*, New York: Holt, Rinehart and Winston, 1973. From the perspective of the present, respective work, neither author goes far enough. In part this is due to their acceptance on some level of the good/evil dichotomy in man; in part, it is due to the lack of a psychobiological etiology of the emotions. Both assume along with their arch-antagonist Lorenz that aggression, albeit benign, is a core drive of the human species. It may be that a better recourse would be to move away from the need to include evil in the description of man's inherent nature. Such a description is possibly due to the influence of certain religions. If one were to posit, rather, that the core emotions of human beings were love of species, which naturally includes love of self and the desire to be loved, this could constitute a first step in the direction of eliminating the concept of evil. It could be that even the love of species is too narrow a designation, for perhaps it should be said that human beings are born with a universal sense of compassion, a sense which extends to all things in the universe without specification. Such compassion of course requires some development and nurture before it can reach its proper expression. In the end, of course, this must be the proper explanation both from the standpoint of the survival of the human species and the survival of the planet and ultimately the universe at large. Human beings (and perhaps other creatures in other solar systems should they exist) are the caretakers of the universe and perhaps, ultimately, of the Divine. There need not be any conflict between a love of species and a love of the universe and the Divine.

3. Spinoza also reaches a similar conclusion: 'Each thing, in so far as it is in itself, endeavors to persevere in its being.' (Prop. VI) and: 'The effort by which each thing endeavors to persevere in its own being is nothing but the actual essence of the thing itself.' (Prop. VII), Part III, 'On the Origin and Nature of the Emotions,' *Ethics*.
4. *Cf.*, Mary Midgley, *Beast and Man*, Sussex: Harvester Press, 1978. *Cf.*, Peter Singer, *The Expanding Circle, Ethics and Sociobiology*, Oxford: Oxford University Press, 1983, p. 5. The existence of altruistic behavior in non-human animals is well documented. Jane Goodall comments: 'In chimpanzee society ... there are examples of individuals risking injury if not their lives to help non-related companions .... One adult male lost his life as he tried to rescue a small infant whose incompetent mother had

allowed it to fall into the water'. *Cf.*, Jane Goodall, *Through a window, thirty years of living with the chimpanzees*, London: Penguin Books, 1990, p. 179. Singer comments on the behavior of elephants who will attempt to raise a fallen fellow elephant to his feet since an elephant will suffocate under its own weight or may overheat in the sun and dolphins who will group under a wounded dolphin to push it upward into the air, for several hours, if necessary. *Op. cit.*, p. 7. The philosopher Wolfgang Köhler reported that a chimpanzee removed a splinter from his finger upon request in the same manner that the chimpanzees doctored each other when the situation so required. *Cf.*, Wolfgang Köhler, *The Mentality of Apes*, New York: Harcourt, Brace and Company, 1931, p. 310. The concept of evolutionary ethics advanced in this present work is not in keeping with Darwin's theory of natural selection based on blind forces selecting some random mutations rather than others. *Op. cit.*, p. 61. The concept of evolutionary ethics advanced here is that human beings can play an active role in guiding the process. This view gains some support from Jane Goodall: '... while chimpanzees will, indeed, respond to the immediate need of a companion in distress, even when this involves risk to themselves, only humans are capable of performing acts of self-sacrifice with full knowledge of the costs that may have to be borne ... A chimp does not have the conceptual ability to become a martyr, offering his life for a cause' (pp. 180-181). Richard Dawkins argues in *The Selfish Gene* that even altruistic seeming behavior is due to the selfish interest of the gene to promote its own survival. Donald Symons argues that genes perform specific tasks such as regulating blood pressure and that 'No mechanism could possibly serve the general function of promoting gene survival because there simply is no general, universally effective way of doing so'. *Cf.*, 'On the Use and Misure of Darwinism', Jerome Barkow, Leda Cosmides, John Tooby, (eds.) *The Adapted Mind*, New York: Oxford University Press, 1992, p. 138. John Tooby and Leda Cosmides argue that genes are themselves the product of evolution and that '... it is the evolved design of the organism that decides what organized consequences the environment can have on it'. *Ibid.*, p. 84. From an Aristotelian point of view, the entity would also be defined by its most mature form of existence (as a tree in its full form is considered to be a tree and a seed is defined in terms of its being a potential tree and not vice-versa). In any event, such an argument appears moot since whether one takes the gene or the individual to be the ultimate arbiter of existence, the result, which includes altruistic behavior, is the same. And, from the point of view being advanced here, human beings in their mature form can and should play a role in evolution regardless of whether this is supposedly being programmed by one's genes. However, even if one analyzes Dawkins' argument on its own terms, the arguments which he advances for his conclusion are insufficient. Particular arguments that he advances are capable of further analysis. For example, in his examination of bird warning calls as purely selfish since thereby a bird can gain the protective cover of the flock, Dawkins seems to rule out the possibility that an act can be both self and species interested simultaneously in the very sense that is being advanced in the present treatment. *Cf.*, Richard Dawkins, *The Selfish Gene* (New Edition), Oxford: Oxford University Press, 1989, pp. 169-170. It is also not implied in this present treatment that all behavior is simply biologically motivated. Bird song may develop for the purposes of aesthetic pleasure. (Hartshorne, an ornithologist in addition to being a metaphysician, has argued in his aptly titled *Born to Sing* that birds sing far more than is necessary for purposes of various forms of communication.) It is only necessary for the present argument that ethics and biology be compatible. It is also important to point out that this present treatment is not a biological reductionism. Higher ethical motivations and theories can develop which are intrinsically valuable. It is only to indicate that ethics serves biology not that ethics serves only biology. It could be argued that a survival of the 'fittest' theory included the necessity of perpetuating altruism because altruistic behavior, in the long run, was a necessary feature of increasing the long-run prospects of the survival of the species. One would do all that one could to keep a human infant alive whose genes showed that the adult human being would develop multiple sclerosis on the grounds that such a human being could grow up to be a Stephen Hawking. With such a perspective, the differences between views of man as being inherently selfish versus inherently good begin to blur. But it is difficult indeed to maintain this thesis as a coherent thesis since it would occasion frequent conflicts as to when selfish and when unselfish behavior was called for. When ethical behavior is qualified by certain criteria, e.g., one should only help those with superior intelligence, other unknown qualities needed for human survival may become extinguished. In any case, one's moral value as a human being may become compromised. One could adopt the standpoint that nature had designed a free-for-all competition between those who were selfish and those who were altruistic such that a certain number of needed altruists would survive. Such a standpoint would require a random distribution of selfish and altruistic human beings after the fashion described by Plato at the end of his *Meno*. Heuristically, such a standpoint would be less effective in terms of achieving cooperative social goals than the thesis presented by the present author. One could also adopt the thesis that the ultimate outcome of universal selfish behavior was the good of the species as a whole. Such a thesis does not account for the survival of altruistic human beings.

5. The term 'borderline' was first used in 1938 by Adolf Stern to describe patients whose symptoms did not fall into the categories of either neurosis or psychosis. *Cf.*, Marsha M. Linehan, *Understanding Borderline Personality Disorder, The Dialectical Approach*, New York and London: Guilford Press, 1995, pp. 4-5.
6. Collingwood, on the other hand, can find no genuine state to which hatred refers. This may indicate that hatred is a spurious concept. He does,

however, acknowledge that fear develops out of love although his argument is far different from the one offered here. For instance, he derives love from hunger, rather than the other way around. His argument suffers from not having considered the overall biological purpose of the emotions. However, one should read Part I of his *The New Leviathan* for it contains much of value though his purpose is different from the present one. *Cf.*, *The New Leviathan*, London: Oxford University Press, 1958, pp. 54-71. There is a growing body of literature on the subject of emotions. However, it appears to suffer from the absence of a deductive hierarchy of the derivative emotions which originate from the master emotion of self-love, the failure to distinguish between ontological states of the emotions, the inability to perceive emotions as part of an emotional continuum, from phenomenological obtuseness, and, with the special exception of Plutchik, a separation of emotions from biology. To choose two problematic examples from Ortony and Turner's article, 'What's Basic about Basic Emotions', the authors argue that fear is a component of distress because '... if a person swimming in shark-infested waters fears an attack by sharks, then he or she will necessarily find that prospect distressing. This means that distress (as we have characterized it) is a component of fear (as we have characterized it), which in turn means that fear cannot be psychologically primitive.' *Cf.*, Andrew Ortony and Terrence Turner, 'What's Basic about Basic Emotions?', *Psychological Review*, Vol. 97, No. 3, 1990, p. 325. But, being distressed about fear does not mean that distress is a component of fear. The distress, which arises, is in reaction to the original feeling of fear and is in no way a component of the fear. The authors continue in the same vein when they state that '... frustration consists of a subset of the attributes of anger' *Ibid.*, p. 327. But, frustration is not a logical subset of anger; it is both an incipient stage of the development of anger and can also arise as a result of the inability to express that anger which can then lead to a different feeling of anger over the consequent frustration. The authors state, however, that '... anger ... we believe is formed by a process of specialization out of the more general emotion of frustration' *Ibid.*, p. 329. One does not phenomenologically begin with a general emotion and specialize from that general emotion. Fear, as it is argued above in this present work, while itself derivative from the inability of self-love to maintain its object is prior to the expression of anger. The absence in the above authors of the proper deductive hierarchy of emotions may be a source of the ensuing confusion. The above authors at first shy away from considering certain emotions basic and when they do, the selection of which emotion is to be considered basic (as shown above) seems to go awry. Ortony, Clore and Collins state that '... we are inclined to reject the idea that there is a set of "basic" emotions such that they, together with their combinations account for all the emotions...'. *Cf.*, Andrew Ortony, Gerald L. Clore and Allan Collins, *The Cognitive Structure of Emotions*, Cambridge, Cambridge University Press, 1988, p. 25. But, 'While we eschew the notion of basic emotions, we do treat some emotions as more basic than others, and we do have a compounding hypothesis for certain emotional states'. *Ibid.*, p. 28. For an approach more compatible with the present one, *Cf.*, R. Plutchik, *The emotions: Facts, theories, and a new model*, New York: Random House, 1962.

7. Hsün-Tzu, 23: 'Human Nature is Evil', Wing-tsit Chan, *Sources of Chinese Tradition*, p. 118. Antonio S. Cua argues that Hsün-Tzu does not mean that human nature is evil as such but that it is neutral stuff and that it is only evil in virtue of the consequences that its self-expression results in strife. While an interesting position, it does not appear to be so advanced by Hsün-Tzu himself and makes Hsün-Tzu resemble Kao Tzu more than himself. In any case, Cua's construction of Hsün-Tzu leaves the basic position of Hsün-Tzu intact which is that human beings are born with envy and hate. This is where Hsün-Tzu, according to the arguments advanced above, has gone astray. Hence, Cua's rescue of Hsün-Tzu, even if accepted, will not affect Hsün-Tzu's starting point. The position advanced here, by altering the starting point, can alter the consequences. Labeling Hsün-Tzu's consequences evil cannot alter them even if that is morality's task to do so. For Cua's reconstruction of Hsün-Tzu, one may be referred to A. S. Cua, 'The conceptual aspect of Hsün Tzu's philosophy of human nature', *Philosophy East and West*, 27, no. 4, October 1977, pp. 377, 383, and, 'The quasi-empirical aspect of Hsün-Tzu's philosophy of human nature', *Philosophy East and West*, 28, no. 1, January 1978, pp. 3, 9, 10. There is some biochemical evidence to support the position that reactions that lead to strife are not even biochemically part of the make up of the original organism, but require a distinct biochemical alteration. The organism, while unthreatened (and thus fearless), would not be prone to violence. Violent acts against another would not be part of the original nature but only part of a disturbed nature. According to Durden-Smith and de Simone:

> The limbic system is said to control emotion and what, as we've said, scientists fondly call "the four F's": feeding, fleeing, fighting and sex. And it directs the body's immediate responses to challenge and danger. When a source of danger is recognized ... the hypothalamus bids up from the pituitary ... a hormone which speeds through the circulation to the adrenal glands with a message to produce cortisol and adrenalin (now usually called epinephrine). These two substances in turn travel through the circulation to the brain and various organs. They arouse the brain and prepare it for fight or flight.

Cf., Jo Durden-Smith and Diane de Simone, *Sex and the Brain*, London: Pan Books, 1983, pp. 116-117. What is suggested in the argument of bio-evolutionary social theory is that human beings, as advanced social beings, have it in their power to socially remove the fear stimuli.

Consequently, the violence response need not ever occur. Social evolution can turn around and influence biology.

8. *Mencius*, II A: 6. *Ibid*., p. 105.
9. An excellent example of the correlation between ethics and biology is the custom aboard ships to save women and children first in the case of a maritime disaster. While this can be considered an ethical act in that one is saving the weaker and the younger when it would be easier for the stronger adult males to save themselves first, at the same time it serves the species since it is more efficient for there to be more females than males and for there to be more children than adults. There is an additional ethical component present in saving the younger first in that the younger have had less time to live and hence are more deserving of more time. The ethical and biological correlation in this maritime custom shows the natural connection that exists between ethics and biology.
10. This is a view put forth by Erik Erikson in his explanation of Gandhi's altruism as a means of expiating unconscious guilt. *Cf*., *Gandhi's Truth on the Origins of Militant Nonviolence*, 1969. One could go further, however, and argue that even if one were expiating guilt feelings (whether conscious or unconscious ones) that this too could be one of nature's devices for prompting ethical action. If a desired and desirable outcome is produced, then even if guilt feelings were one of nature's *upaya* or skillful means, then there is no need to form a cynical view of human nature; e.g., that one is performing good acts for purely selfish reasons. The 'purely selfish' reasons could simply be a ruse for good action - are these then 'purely selfish' reasons? It should also be mentioned that for Mencius it is not necessary that someone act to save the child; it is only asserted that a morally sensitive person would have such a feeling to save the child. The morally sensitive individual might also be a coward and fail to act out of cowardice. It could well be argued that such a failure to act would constitute a lesser degree of ethical development though Mencius does not seem to consider this.
11. Alice Miller has put forth an argument that the cruelty toward and hatred of Jews and even the practise of systematic extermination that was the policy of Adolf Hitler can be traced back to his humiliating early childhood beatings at the hands of his father and the thought that his own grandfather was Jewish. (The systematic extermination is related to the psychological need to eradicate the trace of his own possible Jewish ancestry.) She in no way justifies his grotesque actions and ideas, but considers that even this archetypal case of human evil serves as an illustration that evil is not a gratuitous tendency randomly instilled in one's nature with birth. *Cf*., Alice Miller, *For Your Own Good, Hidden cruelty in child-rearing and the roots of violence*, Hildegarde and Hunter Hannum (trans.), New York: Farrar, Straus and Giroux, 1985, p. xi and especially, 'Adolf Hitler's Childhood: From Hidden to Manifest Horror,' pp. 142-197. Hans Küng points out the religious roots of Hitler's anti-Semitism, first learned as a six-year-old choirboy who served at mass. *Cf*.,

Hans Küng, *Judaism*, Munich: Piper Verlag, 1991, pp. 236-7. For a very different point of view, one may be referred to James Hillman, *The Soul's Code*, New York: Random House, 1996, Chapter Ten, 'The Bad Seed'. In Hillman's wry prose, quoting Alice Miller, ' "Hitler actually succeeded in transferring the trauma of his family life onto the entire German nation" ... It sounds as if the whole course of world history could have been altered by early therapeutic intervention in that obscure Austrian household. Twenty million Russian casualties and six million Jews, not to mention the victims from all the other places, and the dead Germans, too, caused by young Adolf's beatings and the behavior of his mother, etc.', p. 229. But Alice Miller might respond, 'Precisely'. Ron Rosenbaum criticizes Miller's argument in his *Explaining Hitler, The Search For The Origins of His Evil*, New York: Random House, 1998, pp. xxi-xxii. A wider and a different thesis is provided by Horkheimer and Adorno in their *Dialectic of Enlightenment* in which they find, as does Freud, the origin of Germany's descent into barbarism the result of an inordinate level of instinctual repression. While this is the essential insight of Freud's *Civilization and its Discontents*, it was Walter Benjamin who first wove this into a philosophy of history. *Cf.*, Richard Wolin, *Walter Benjamin, An Aesthetic of Redemption*, New York: Columbia University Press, 1982, pp. 266-269. It was the work of the enigmatic Walter Benjamin that set the stage for the seminal work of the Frankfurt school which opposed the view of the Enlightenment as articulated by Condorcet that historical development was positive and progressive and proposed instead the concept of negative dialectics, i.e., that historical progress led to renewed barbarism. Benjamin's thesis was more idiosyncratic in that for him, the ruptures of history, not its directionality, became his chief point of focus. (The present, respective author had the inestimable fortune not only of meeting Herbert Marcuse but also for a time being a working colleague of Herta Jogland, both 'members of the Frankfurt school'.)

12. This does not mean that it serves only biological drives but rather that it is not in conflict with biological drives. This view should be distinguished from the general outlook expressed by those who have identified themselves as socio-biologists. E. O. Wilson, as a prime example, seems to interpret biological drives as somehow in conflict with higher evolutionary ideals and approaches such as social altruism. Consider, Wilson's caricature of saints in general:

> Sainthood is not so much the hypertrophy of human altruism as its ossification. It is cheerfully subordinate to the biological imperatives above which it is supposed to rise.

Cf., E.O. Wilson, *Sociobiology: On Human Nature*, Cambridge: Harvard University Press, 1978, p. 166. Michael Ruse in his essay, 'Morality of the Gene' quotes this passage from Wilson as indicative of his point of view in general in which such a figure as Mother Teresa would arise in society in conflict with the biological drive towards survival. *Cf.*, Michael Ruse, 'Morality of the Gene', *The Monist* (Vol. 67:2, April, 1984), p. 172. The problem with the socio-biologists, as distinct from the bio-evolutionary-social model, which is proposed herein, is that biology is seen as a model for human behavior without taking into account its modification, in fact its realization, in social evolution. While Wilson is of course an extreme example of this type of thinking, on the other hand, the consideration of biology abstracted from social evolution seems to be fairly characteristic of this group as a whole. This understanding of biology would appear to exempt the brain and/or the mind whereby the species could become aware and in fact must become aware of more effective modes of species survival if it is to survive at all.

13. In a manner of speaking, it may be said that the bio-deduction of the emotions in a systematic sequence from basic emotions which are in turn a manifestation of psychological drives which reflect basic metaphysical characteristics, may be compared to Kant's self-description of his arrangement of his categories which he writes in his letter to Herz of February 21, 1772 were placed side by side by Aristotle, in contrast, in a purely chance juxtaposition. *Cf.*, Arnulf Zweig, (ed. and trans.) *Kant, Philosophical Correspondence 1759-99*, Chicago: The University of Chicago Press, 1967, p. 73. The present author's systematic derivation and hierarchical ordering of the emotions in a causal sequence from basic to derivative emotions is offered as a contribution to the discipline of psychology.
14. Chung-ying Cheng, 'Chinese Philosophy: A Characterization', *Invitation to Chinese Philosophy*, p. 163, *et passim*. While nowhere herein does Professor Cheng explicitly state that it is an insight of Chinese philosophy that man is not anti-social by nature, it seems that he would consider social harmony to be a subset of harmony in general. Wing-tsit Chan also seems to express this view: '... the foundation of the Confucian system lies in the moral realm, that is, in human experience itself. The thread is also generally taken to be identical with the Confucian doctrine of central harmony (*chung yung*, Golden Mean). Indeed, this doctrine is of supreme importance in Chinese philosophy; it is not only the backbone of Confucianism, both ancient and modern, but also of Chinese philosophy as a whole. Confucius said that " to be central (*chung*) [with all]" is the supreme-attainment in our moral life. This seems to suggest that Confucius had as the basis of his ethics something psychological or metaphysical'. *Cf.*, Wing-tsit Chan, 'The Story of Chinese Philosophy', in Charles A. Moore, (ed.), *The Chinese Mind, Essentials of Chinese Philosophy and Culture*, Honolulu: East-West Center Press, University of Hawaii Press, 1967, p. 35. *Cf.*, also Robert E. Allinson, 'The Confucian Golden Rule: A Negative Formulation', *Journal of Chinese Philosophy*, Vol. 12, No. 3, September 1985, pp. 304-315.

Largo maestoso
Slowly and majestically

2 The Moral Realm of Truth and Mencius' Phenomenology of Compassion

Each philosophical realm possesses its own unique kind of truth peculiar to it. The moral realm is no different from any other philosophical realm in this respect. While it is true that Plato reveals the failure of Socrates to persuade his antagonists such as Thrasymachus or Callicles of the truth or the value of ethics, it does not follow from this inability to persuade others that ethics does not belong to the realm of truth.

For some unaccountable reason, however, the twentieth-century analytic tradition of Western philosophy understood Socrates' failure to persuade Thrasymachus and Callicles to be an indicator that ethics belonged to a different realm than that of mathematics and perhaps the rest of philosophy, and, whereas it was perhaps possible to reach certainty of truth in mathematics, such certainty was unavailable in the ethical realm. Carried through to the extreme of its logical conclusions, ethics in the twentieth century degenerated, in terms of leading Anglo-American thinkers, into ethical emotivism in which it was held that no truth at all belonged to ethical claims, but only how individuals felt. Since feelings were perceived by these thinkers as subjective and arbitrary, the collapse of ethical claims into emotions provided further evidence of their unreliable and relative character.

Ethical truth has just as much claim upon truth as aesthetic, religious, factual, or logical truth. A factual truth cannot be true unless it is possible that there is truth, for what would it mean that something were true, unless one were to understand the meaning of truth. This is easier to see in the case of the Beautiful. It is possible, though slightly difficult, to amass a fairly universal agreement that Mozart's music is beautiful. Now this must mean that Objective Beauty exists since the beauty of Mozart's music is not confined to any one listener's private sphere of enjoyment. However, if one claims that it is wrong to kill a person, the cries of cultural, contextual and historical relativism will very quickly surface.

How does one know that the truth that murder is morally wrong is a certain and universal truth? One knows this in no less certain terms and in no different way than when one knows that two straight lines can never enclose a space. Whenever one hears the proposition, 'murder is wrong', one knows with an infallible certitude that this is necessarily true. It is not merely that one agrees with the proposition; it is that one cannot help but agree. The agreement with the truth of the proposition is derived from its truth; its truth does not derive from its nearly universal consent. Its truth cannot derive from its nearly universal assent, because one knows its truth before one needs to collect or collects the data of consent.

The truth of the proposition, 'murder is morally wrong', is not derived from the universal or near universal assent given to this proposition whether intellectual or emotional. The nearly universal assent given to this proposition derives from its truth. The 'nearly' derives from qualifications of this truth. One may kill, so it is thought by many, in military operations in times of war, in self-defense, in selected cases of euthanasia, in selected cases of abortion, and as capital punishment. Whether or not one agrees with all of these exceptions, the point being made is that these are qualifications concerning the strict universality of the extension of the proposition. These qualifications are qualifications concerning the circumstantial applicability of the necessarily true proposition; they do not derogate from the truth quality of the proposition. In these exceptions, it not that it is right to kill; it is that another moral claim assumes a higher ethical priority to the prohibition against killing. For example, in the case of euthanasia, the moral injunction to alleviate endless severe and irreversible suffering may supersede the moral injunction not to kill.

One must also consider the case of the morally senseless. The inability to assent to the truth of this proposition is a proof of moral vacuity. Such an inability to assent to the truth of this proposition or even the active disagreement with the truth of the proposition, as evidenced by the Nazis discussed in the previous chapter, is a mark of ethical degradation or the defining characteristic of the lack of ethics or morality. It is always wrong to commit murder. It may be impossible, at times, to convince an immoral person or group of people of this truth but such a failure of rhetoric does not make it less true.

When one hears the phrase, 'murder is morally wrong', a morally sensitive agent (one who has passed Mencius' test) knows, upon hearing this phrase, that it is absolutely and certainly true. The very absoluteness of its truth and the certainty of its truth derive from the fact that it belongs to the order of truth. The ethical order of truth is a division of the order of truth. That Mozart's music is beautiful is also a certain and universal truth.[1] What is beautiful is beautiful because of its truth or correspondence with the highest realm of aesthetic order or harmony. The beautiful is beautiful, for example, because of its incorporation of good and just elements of reality or being: the beautiful is the result of the proper proportion of the elements of unity, symmetry, sensitivity, delicacy, spontaneity, passion, redemption, affirmation, acceptance, tragedy, whimsicalness, imaginativeness, finitude, eternity, transcendence, immanence, meaningfulness, truth, goodness, character, justice, fatefulness, order, measure and harmony.

'Thou Shalt Not Kill' is an injunction that reality, life, being, truth and beauty should not be destroyed or diminished. The reason that this is known and felt to be true is that any diminishing or destruction or defacing of reality is wrong. Reality possesses its own right to be. Reality or existence contains its own imperative within itself. This imperative, its right to be, its right to live, is as absolute as its claim to be beautiful. The necessity of the truth of the proposition, 'murder is morally wrong', derives from its correspondence with the existential claim of reality. Every Truth is at once a reflection of Being, of Beauty and of Reality. Every act of unnecessary violence, of thoughtlessness, of gratuitous harm doing, is a senseless mark of *graffiti* on the face of the Divine.

Up until this point, the argument has been that the statement 'murder is morally wrong' is a necessary truth and can be recognized as such. One need not and in fact cannot conduct an empirical survey to collect opinions on this subject. Even if one were to conduct an empirical survey, the results of the survey would not at all affect the truth of the proposition.

However, in the case of moral truths, unlike mathematical ones, one can also recognize the truth of the moral claim by a feeling of compassion that is aroused by the witnessing or imagining of morally relevant data.[2] Chinese philosophy is renowned among philosophies of the world for its ethics.[3] Mencius, a Confucian philosopher, perhaps the greatest of all philosophers of ethics in ancient China, who lived in the 4th C., B.C.E., devised a phenomenological test for the unsolicited and disinterested arising of compassion as a means for discovering the absoluteness of the moral realm of truth. While Mencius did not, of course, describe his discovery in these modern terms, such a description is fully compatible with his ethical project.

Mencius did not so much as argue as present a phenomenological test case as a thought experiment so that all human beings could discover that the ethical feeling of compassion was necessarily aroused in the observer in cases where the human life of an innocent other was in imminent peril. Mencius' test case example, designed as a thought experiment, is famous in Chinese ethics, and is known as the example of the child in the well, which was examined in the previous chapter. Mencius' methodology reminds one of Kant's famous use of examples such as 7 plus 5 equals 12 to establish epistemological claims. Of course, such an example can be updated to that of a child who has run out into the middle of the street in the path of an oncoming car. The point of the example is that, according to Mencius, all human beings will of necessity be moved to compassion at the prospect of the imminent loss of innocent human life. Mencius does not argue that all men will take action to rush in and save the child. That would be to conflate ethical consciousness (compassion) with ethical behavior (action). All he needs to make his case is that all men will of necessity be immediately moved to compassion. Mencius wishes to show that there are no extrinsic reasons for morality. The value of innocent human life is ultimate. There are no exceptions to this. Whatever the circumstances, one must feel compassion for the innocent one who is in danger of losing her or his life. The feeling of compassion that is aroused is neither for the sake of appearances nor because of any reward that the observer hopes to win. *The feeling of compassion is necessarily aroused at the sight of the imminent prospect of the loss of innocent human life.*

It may be objected that this is an incomplete account of morality since compassion does not necessarily lead to ethical action. Such an objection is essentially sound because it points up the limitations of compassion if compassion is understood as sufficient to account for a complete account of morality. One would need to appeal to motivations other than compassion to ensure that one's ethical feelings would issue in ethical behavior. One would need to appeal to a concept of just action or benevolent action in order to offer a more complete account or morality.

There are other limitations to compassion as a complete account of morality. One should feel no compassion or extremely limited compassion for the murderer, for example. If one were to feel too much compassion for the murderer, the murderer (and society at large) might well derive the impression that murder was sanctioned. The result of this sanction might be the toleration of further acts of murder by this murderer or other murders. Thus, compassion must be kept in check by a careful consideration of who is selected as the recipient of the compassion. In Mencius' case this problem is well avoided since the example is constructed so that an innocent child rather than a murderer is the recipient of the compassion. Of course, there is the omnipossibility that the child could grow up to become a murderer, but this assumes a life which is determined in its outcome.

Nevertheless, the above consideration once again emphasizes an important concern. Compassion must be well specified in respect of who is deserving and who is most deserving of compassion. Compassion must not be felt, or must be felt in only a very minimal degree for the harm-doer. (Of course one has in mind here the cruel murderer, not the starving thief.) It requires only a moment's thought to realize that the handicapped, the infirm, the ill, the abused, the unloved, the aged, the poor, orphans, small children, those who are wrongly prejudiced against, those who are victims of social injustice, those who are in mourning, those who have lost hope, those who are isolated, those who are lonely, those who are neglected by others, the stranger and the widow would be the most deserving of compassion. When one couples the right exercise of compassion with the need of doing good deeds in order to ensure that compassion finds an appropriate outlet, what qualifies as an ethical action becomes more clear. For example, an ethical duty would be visiting the sick. Another would be showing hospitality to strangers. Another would be offering the gift of companionship to the lonely. Yet another would be comforting those who are grieving. Many duties follow if one considers that one's prime duty is to relieve the suffering in the world. Mencius' example is only the beginning. The point of this example is that an immediate argument has been offered that life is sacred. One does not need to reach this conclusion only by contemplating the truth of the proposition that murder is wrong. One can reach the conclusion that life is sacred in the immediacy of the circumstances in which one is faced with the prospects of witnessing the extinction of human life.

Now, this is not an empirical theory even though an empirical example has been offered. If one has recourse to Karl Popper's criterion of non-falsifiability, then this "theory" being non-falsifiable, is also without any truth-value. As a result, it is important to realize that the example is not a means of seeking empirical confirmation or disconfirmation; it is only a means of illustrating a necessary truth.

Mencius has produced a phenomenological test for the existence of compassion as a basis for ethical consciousness. It is not only a phenomenological test. It is a test that shows that for the person with an ethical consciousness, ethical feeling or compassion is necessarily aroused by the sight of suffering without any further motivation being needed. The desire to remove suffering is not for the sake of any other reason. The fact that it arises spontaneously, categorically and with certainty is a proof that ethical consciousness is both absolute and unconditioned. It is not based on considerations of utility or gain or happiness. The phenomenology of compassion - in Mencius' brilliant example - is a proof of the moral realm of truth. If one is willing to perform Mencius' feeling/thought experiment in one's imagination, one then possesses a key to knowing that ethical consciousness is known by the individual subject knower as an absolute truth and no further evidence can controvert this experience of truth.

From an epistemological standpoint, the knowledge of ethical truth is exactly similar, in the following respect, to the knowledge of mathematical truth. Once the ethical truth is known to be true, it is known to be always true. There is no more need to repeat the thought experiment of the child in the well any more than there is a need to continue to add up one and one to make sure that the sum is still two. The truths of mathematics do not depend upon repeated empirical testing. Ethical truths also do not depend for their truth-value on repeated empirical tests. Once an ethical truth is known to be true, it can never be false. Ethical truths are known to be necessarily true at the moment that they are known to be true.

The knowledge of ethical truth bears a similarity to the knowledge of mathematical truth in another sense as well. Ethical truths do not depend upon outside corroboration for their truth-value. One does not need others to perform Mencius' experiment to confirm or disconfirm the personal discovery of ethical consciousness that each individual can in fact, nay, must make for herself or himself. In one's self-knowledge of the truth of Mencius' test case, one knows that one's knowledge is universal. But, the universal knowledge that one gains by subjectively imagining oneself to be in the position of the observer at the well is a universal knowledge that each subject must discover for herself or himself. The knowledge of ethical truths is only acquired through personal experience. That others may be found who also arrive at ethical knowledge via their own phenomenological experience of Mencius' test is an illustration of the universality and necessity of Mencius' claim, but it is not a confirmation of Mencius' claim. Every instance of someone who recognizes the ethical truth that is contained in

Mencius' example constitutes a new discovery of that person and for that person of the ethical truth. It does not further confirm the truth; it simply serves as an additional self-discovery and illustration of the truth.

It is also important to note that the truth of Mencius' example of necessary compassion is not derived from the child in the well example. In a sense exactly similar to the case of mathematical truth, the morally relevant example only serves to illustrate that truth - in this case moral truth - must be illustrated in experience, but its truth-value cannot arise from the experience. If the truth-value of the moral truth arose from the experience, Mencius' theory would be an empirical one. But, Mencius is not proposing a collection of empirical, morally relevant data. In fact, it cannot be an empirical collection of morally relevant data, since just one experience or subjective re-enactment is enough to illustrate the truth that compassion is an absolute. One does not need to keep standing by the well while children are brought by to be held hovering at the edge of the well in order to keep testing the hypothesis of compassion and alarm. This would be equivalent to continuously adding up one and one to make sure that the sum is still two. One may adapt what Kant has said about knowledge: while all truths must begin with experience, not all truths arise out of that experience.

The suddenness of the feeling of compassion is suggestive of something else as well. It is suggestive that moral truth is not something that is invented. Moral truth exists to be discovered. It is interesting that in the case of the history of Western ethics, very few test cases can be found which test the universality and necessity of the ethical consciousness. In Plato, with the example of the ring of Gyges, there is no question of an instant action which is called for; it is a case of deliberate action which can take subjective benefits into account.[4] The example differs so radically from that of Mencius that it cannot be fruitfully compared. Plato's example of Leontius not being able to tear one's eyes away from the sight of dead bodies is a closer example in that it takes into account one's sudden and unmotivated course of action:[5]

> But, I said, I once heard a story which I believe, that Leontius, the son of Aglaion, on his way up from Piraeus under the outer side of the northern wall, becoming aware of the dead bodies that lay at the place of public execution, at the same time felt a desire to see them and a repugnance and aversion, and that for a time he resisted and veiled his head, but overpowered in despite of all by

> his desire, with wide staring eyes, he rushed up to the corpses and cried, There, ye wretches, take your fill of the fine spectacle.[6]

But, this example does not involve any choice that would either harm or help a human being (since the test case is of corpses), and, as a result, does not provide any information about how a human being would feel in a suddenly demanding situation which involves the probable and immediate loss of another human life.

With Mencius' case of the young child about to fall into the well, the likelihood that occurs to the observer is that another human being, a helpless human being, is about to die. In one's subjective re-enactment of this thought experiment, one must imaginatively experience the child about to fall into the well and drown. This is the point of the example which is to be experienced. One must put oneself in the shoes of the observer who is about to see the child helplessly and ineluctably fall into the well to a certain death by drowning. Under such a circumstance, whether in the real order or in the order of the imaginative reconstruction, compassion is immediately aroused as a universal reaction. Ethical consciousness, for Mencius, is immediately engaged whenever circumstances present themselves wherein innocent human life is mortally and imminently threatened.

Of course, the further question, which this example of Mencius' does not fully address is, if one does not possess such a feeling, if one possesses no ethical consciousness, then how can one be convinced that one should possess an ethical consciousness? One may claim that one neither intellectually recognizes the truth of moral claims nor does one possess any feelings of compassion. If one claims that one possesses no intellectual recognition of moral truth, Mencius' example provides an avenue for the recognition of moral truth. But, if one claims that one possesses no moral feeling in the real or imagined case of Mencius' child in the well, then how can one be persuaded to be moral? Mencius does attempt, elsewhere in his writings, to account for exceptions by reference to negative conditioning and a lack of education. While a full discussion of Mencius' moral theory is beyond the limits of this work, the bottom line is that there can always be someone who possesses no compassionate feelings, edits out compassionate feelings or otherwise refuses to heed them. How one can persuade others to be ethical is a classical problem for ethics, and Mencius cannot answer this question any more than any other ethical theorist. But, Mencius cannot be faulted for not providing an answer to this question as shall be seen below. What must be kept in mind is that the presence of compassion is an essential part of what is meant by the possession of an ethical consciousness. If compassion is absent and one possesses no feelings of pain at the sufferings of others, such an existence is definitive of a cruel and/or indifferent being. (A being may be cruel and not indifferent such as those who take pleasure in the suffering of others as the word *Schadenfreude* which exists in the German language describes.) The cruel or the indifferent being is the unethical being. The opposite of compassion is cruelty and/or indifference to others.

If one were to persist in raising the question, why actualize ethical consciousness, there would be no answer to this question. It must be the case that every human being is obliged to possess a moral consciousness, and this obligation is absolute. One's obligation to be moral is an ultimate obligation, and cannot be justified or rationalized in terms of higher obligations and higher values.

The question then becomes, whence comes the obligation to be moral? It may be that ethical action or the abhorrence of unethical action springs from human compassion, but the directive to listen to that compassion must spring from the knowledge that one's moral commitment is an ultimate commitment. Mencius' example may show that there is a moral realm of truth, a realm in which one's ethical consciousness is necessarily aroused with no thought of an external reward, but what proof can be given that one should heed such an ethical consciousness? Is it possible to prove that one should be moral?

It can be argued that in order to become a fully developed human being, one must develop one's physical, intellectual, religious, aesthetic and ethical consciousness. One does not achieve ethical consciousness simply by blindly following an externally dictated law of behavior. Following externally dictated laws of behavior might possibly create good results, but it does not create ethical consciousness. In order to become a fully developed human being, it can be argued that one possesses an obligation to develop an ethical *consciousness*. Nonetheless, one can always raise the question, what necessity is there in becoming a fully developed human being with an ethical consciousness? As Raskolnikov wonders in Dostoyevsky's *Crime and Punishment*, 'Why should I be moral?'.

It is difficult, if not impossible, to give an answer to the question, why should one be moral? Just because one has no answer to this question, however, it does not follow that one need not be moral. The only answer that can be given to this question is this one: human beings must be moral; there is no further reason why, nor should there be. It does not lessen the rightness of the demand that one should be moral that one cannot supply a reason why.[7]

One might ask, why be aesthetic or why be intellectual or why be religious? One can exist without a love of art or music and without a critical mind and without a feeling of reverence towards the Divine, nature or the cosmos. One can also exist without an ethical consciousness or ethical behavior. But, the existence of a human being who lacks an ethical consciousness is more unacceptable than the existence of a human being who lacks an aesthetic or an intellectual consciousness. This fact by itself indicates that there is a stronger injunction to develop a moral consciousness than there is to develop, for example, an intellectual consciousness or an aesthetic consciousness. One does not consider that a construction worker must be capable of delighting in Bach or critically reviewing an interpretation of Hegel. But one expects that a construction worker must be capable of an ethical consciousness and ethical behavior.

If a human being exists without a love of art or music, poetry or drama, dance or opera, literature or philosophy, then what sort of human being exists? Surely, one would not think highly of such an existence, but would feel only pity for such a being if capacities for such enjoyment existed but were unactualized, or even contempt if opportunities for such learning existed, but were rejected.

If a human being exists without a sense of questioning of her or his place in the universe, without wishing to inquire further into the laws of science or medicine, without desiring to learn about the history of civilizations, without any desire to critically evaluate laws which she or he is commanded to obey, then what sort of human being exists?

If a human being exists without a sense of gratitude for being alive, without a feeling of compassion for all beings that suffer, without a sense of mystery and wonder at the possibility of life, without a deep feeling of sadness and hopelessness and outrage over the sufferings man has inflicted upon his or her fellow man, without a wish to make things better in the future, then what sort of human being exists?

If a human being exists who willfully hurts or kills other human beings and who feels no regret over such action, the question is not merely what sort of human being exists, but how can one stop such a human being from killing other human beings and how can such a human being be punished for her or his actions? One does not contemplate punishing the aesthetic dolt or the intellectual dullard, but in the case of the perpetrator of moral evil, one considers and even demands that there be some kind of redress.

Why be moral? The question seems to presuppose that one has a luxury of choice along the lines of why play the piano, or why take up cooking? It is not that one cannot supply an adequate reason, and, therefore the choice to be moral is arbitrary. The choice to be moral or not to be moral is itself a morally accountable choice. If one chooses not to be moral, that is an immoral choice. Or, if one chooses to be amoral, one is also making an immoral choice. Assuming sufficient capacity, sanity, maturity, awareness, education, and intelligence, one cannot avoid making a moral or an immoral choice. Moral responsibility is an inescapable necessity for human beings. One can, of course, choose to be immoral or amoral, but such a choice is nonetheless unavoidably laden with moral responsibility.[8]

Of course, one can offer reasons for morality, but no reasons are adequate since morality is already an ultimate commitment. If a reason or reasons could be given for morality, then *that reason* or those reasons would constitute one's ultimate commitment, and that reason could then be attacked for arbitrariness. The question would simply be pushed one step further back, and there is no reason why this would not become an infinite process or an infinite regress of demands for justifications for one's justifications. The cry for endless justification is itself an indictment of reason. Endless justification simply shows the futility and the utter inadequacy of the instrument. Reason and justification are tools, which are or should be in the service of higher values, which ultimately can only be self-justifying.

Suppose, for example, one were to say that one's ethical actions were justified, because they were religious commandments. In that case, one's ultimate commitment is to one's religion. There is nothing wrong with that. So long as one's religion is not the religion of the early Incas that demanded the sacrifice of infants, or the Chinese religion of the Chou period that required human sacrifice, or the early Viking religion that demanded both infant sacrifice, mass rape and thereafter dismemberment of widows, there is nothing to criticize if one's ultimate commitment is to one's religion. A worthy religion must be one which commands respect, which is founded on values of love and compassion, which, in a word, meets the test of morality.

Normally, one does not consider that there is something wrong with not being able to justify one's religious commitments. One does not consider one's commitments thereby to be arbitrary and unimportant. Why then must morality shoulder such a burden? One can, of course, offer reasons on morality's behalf, but such reasons cannot be ultimate justifications or morality would not be an ultimate call to commitment. One can argue, for example, that one should not deceive others, because society would be impossible if everyone engaged in deception. But, while such a reason is and should be fully compatible with morality, it does not justify morality.

The question, why be moral, is in fact a pseudo-question. It is a pseudo-question, because it is couched in a grammatically correct interrogative form, and thus carries the linguistic appearance of being a genuine question, while it is not, because morality does not need and cannot possess an extrinsic reason for its being.

Morality serves life, but it serves its own higher ends as well. With morality, all human beings may live in a state of absence of fear and anxiety when they travel in the presence of strangers. Morality is the lubricant of human interactions. One may feel confident when one mails a letter that the postman will deliver it instead of throwing it aside. Without morality, one would live in a constant state of distrust, suspicion, discontent and paranoia. But, the fact that morality secures these ends is not the reason for its being. Morality provides a higher meaning for human life, just as the love of beauty, harmony and truth provides a meaning for life that transcends the mundane routine of existence. But, morality does not exist to provide a higher meaning for life. It is because the demands of morality are transcendent, that they cannot be reduced to serving other ends, that they can provide a meaning for life. If moral values existed solely for the sake of providing meaning for human beings, perhaps they could be replaced one day by other meaning-producing devices. *It is only because the demands and the truths of morality are irreducible that they give meaning to life.* If they are reduced to being meaning-producing devices, they will cease to give meaning to life, for the purely useful and functional can never serve as a source of transcendent value and ultimate commitment. Indeed, if human beings prove not to be the only source of higher conscious life in the universe, morality justified in terms of serving human life will appear anthropocentric in the extreme. Moral values must be transcendent to purely human values both in order to serve as transcendent and thereby adequate sources of obligation and as adequate guidelines to universal conduct.

The inability to offer a reason to be moral should not be a source of embarrassment to philosophers. If a reason could be offered, then morality would be an inferior instrument, which existed only for the sake of serving the higher reason. That no reason can be given is not a proof of morality's lack of foundation. The demand for a reason for morality is a proof of a false religion: the deification of reason, of reason giving or of justification. If one is called upon to supply a reason for everything as if reasoning were the only reliable court of appeal, then reason would be the god which everyone is being commanded to follow.

It would be more embarrassing if one could offer a proof as to why one should be moral than if one could not offer such a proof. If a proof could be given for why one should be moral, then one would be 'forced by logic', as it were, to be moral. If one were compelled by some external proof to be moral, then what kind of commitment would morality inspire? One would be a kind of robot submitting to the power of some kind of proof. Where would one's freedom lie if one's morality were simply a response to some external obligation? Why should one praise moral action if one had no choice but to follow the proof that one should be moral? It would be as if one were to praise the child that attended compulsory lessons, because she or he were forced to do so by her or his parents, more than to praise the child that voluntarily out of a belief that education was a high value, attended lessons that she or he was not forced to attend.

No, one should be more worried if one could prove that morality had a reason to exist, than if one could not offer such a proof. If no such proof can be given, then morality is a matter of pure choice and freedom of will. Then she or he who chooses to be moral is to be praised, and such behavior is to be prized. *The Book of Job* provides eloquent testimony that no reason can be given for why one should be moral. This is its message, that morality is beyond any price, for any price put on its head would take away from its freedom and thereby its ultimacy of choice. It could be argued that more than absolute faith, absoluteness of moral choice is the message of Job. Despite all that happened to him, and throughout all of his suffering, Job did not lose his integrity.[9] He also retained his faith in the Divine. The story of Job can certainly be interpreted as an object lesson in faith. However, from the standpoint of morality, Job's decision to retain his faith was also an act of moral integrity. Morality for Job was not to be derogated. It was an ultimate value, as was his faith in the Divine.

The moral point of view is precisely the point of view that in freedom a choice is made for the good or the right. In choosing the good or the right a higher freedom is thereby won. In fact, it is the moral value, the rightness of the choice, that informs the will, that gives the otherwise blind will content and direction. The moral value itself is paramount. The absoluteness of the choice of the good over the evil is what bestows true and complete freedom to the will. When forced with the choice, only the person who chooses to suffer evil rather than to commit it is truly free.

If moral values are not transcendent, then they cannot command this order of allegiance. It is only if moral values are transcendent that they can act as a sufficient guide to the freedom of the will. Without values in general, freedom lacks content, order and direction. Values give guidance and order to the will. Moral values give to the will its highest freedom. It is only if moral values are valued for their own sake that freedom is thereby achieved. One cannot worship at the shrine of abstract and empty freedom.

One way of understanding the absoluteness and the necessity of moral values is to construe moral values as part of a Divine ontological order. It is difficult to be inspired by abstract higher values alone. The pursuit of higher values in general and moral values in particular can serve as an inspiration if such a pursuit is construed to imitate the Reality of the Divine and to reflect the Divine order.

Whatever is the necessary ontological explanation of moral values, moral values always retain their own unique form of absoluteness. Socrates' willingness to suffer evil rather than commit it loses none of its moral absoluteness. That it is morally wrong to gratuitously kill one's fellow human beings is an eternal, necessary truth that can never be otherwise.

If one acts morally, because one is following some sort of proof that one should act morally, then one's higher commitment is to proof rather than to morality. One's higher commitment is to the means, and not to the end which the means serves or should so serve. One's higher commitment is to being forced (by having it proved) to act, to be compelled to act by the compulsory power of proof, rather than choosing to act in the absence of any other reason than that the moral value that one should not gratuitously kill or otherwise gratuitously inflict harm upon another human being is sufficient reason to be moral. Can or should any other reason supersede this latter motivation? If any other reason were to supersede a moral value itself as a motivation for action, then morality would no longer be an ultimate value.

Morality not only cannot be proved by something higher, but it should not be so proved. If one does not realize – to take the nearest example - that no human being should be gratuitously murdered, then one cannot be made to understand this through any form of proof. But, the inability to prove that one should be moral does not mean that a commitment to not killing others is meaningless. The unanswerability of the question, why be moral?, only means that fatuous questions can be raised which are not deserving of answers. The unanswerability of the question, why be moral?, is itself a condition of the ultimacy of moral commitment. If the question, why be moral?, could be answered, then morality would no longer be an ultimate commitment and moral values would no longer be ultimate values.

The demand to provide a proof for morality is an intellectually decadent demand. Such a demand is sophomoric. Every question cannot be answered. Some questions do not deserve even to exist as questions.

The question, why be moral?, belongs to this class of questions. Either one must be capable of intellectually recognizing moral truth, or one must be capable of passing Mencius' test of moral compassion.

If one does not immediately feel sympathy for and desire to respond to the hungry in the world, then there is something wrong with that person. If that person demands a proof before action or considers that ethical action is arbitrary, and therefore can wait until some other arbitrary desires are fulfilled, then there is something lacking in that person. The very demand for a proof of why one should be moral is strong evidence of moral numbness and moral degeneration. If one is led to think that the question, why be moral?, is a legitimate question to answer, then one has lost one's moral senses. In the time that one has taken to endeavor fruitlessly to answer this question, many lives could have been saved from hunger, illness and injustice.

But, it is a sign of the times that not only that such a question can be asked, but that it is considered that such a question is worthy of being answered. No proof for why one should be moral can ever, or should ever, be given. The existence of the moral consciousness is an existence, which recognizes its own reason for existence - it supplies its own reason for existence out of itself. It supplies a life experience of the moral value that with no thought of benefit to one, one's immediate value is to feel compassion for the other. Mencius' example is designed as a phenomenological method for recognizing the existence of moral truth; it is not designed as a proof that one should be moral, although it can be understood to have the power of exhortation. If one understands Mencius' example aright, when one feels compassion for the child who is about to die, one also realizes, by extension, that an ethical life consists in the attempt to alleviate the suffering of others. If one needs further encouragement to be moral, one can point to the existence of heroes and heroines who risked and paid with their lives to save others, ordinary people such as one's neighbor, Dr. Johnson, who gives of himself to charitable work, or saints and sages, both sung and unsung, whose exemplary lives are spent in alleviating the suffering of others.

Notes

1. This is said despite the late Isaiah Berlin's statement, in his account of Tolstoi, that 'We find the works of Mozart and Chopin beautiful only because Mozart and Chopin were themselves children of our decadent culture, and therefore their works speak to our diseased minds; but what right have we to infect others, to make them as corrupt as ourselves'. *Cf.*, 'Tolstoi and Enlightenment', *Russian Thinkers*, Henry Hardy and Aileen

Kelly, (eds.), Harmondsworth: Penguin Books, 1979, p. 255. It is entirely possible, though it is not clear from the essay, that Sir Isaiah, normally a most perspicacious thinker, was merely restating Tolstoi's view. Of course, it might be that Sir Isaiah's view (or Tolstoi's) would find support from the example of the feared Josef Mengele, the worst of the death-camp doctors in Nazi Germany, who cruelly experimented on inmates, despite the fact that he was well educated, had studied Dante, and was a lover of music. The innocent children called Mengele 'uncle'. He gave them chocolates before murdering them. Perhaps, Rousseau was right after all. Of course, with a wider, Confucian definition of culture, one who had not attained to morality would not be considered a cultured individual.

2. According to Kupperman, sympathy is not a reliable guide to ethical action: 'Someone can be very sensitive to, and sympathetic with, the feelings of everyone affected by her or his actions, and still make the wrong choices.' *Cf.*, Joel J. Kupperman, *The Foundations of Morality*, London: Allen & Unwin, 1983, pp. 62-3. In the case of Mencius, it is compassion and not sympathy that is the hallmark of ethics. One is tempted to answer that one cannot be persuaded to make the wrong choices since one's compassion is for human life and human suffering, not sympathy for feelings in general and thus one would be protective of human life and not of various feelings that may be aroused. On the other hand, Kupperman's fear that one may make the wrong choices is a valid fear even if one's compassion is for human life if one's compassion is for the wrong human life. For example, if one were so warped as to attempt to protect the murderer out of compassion for the murderer instead of the murderer's intended victim, one would thereby be misled by compassion by having compassion for the wrong person. This is an odd case since compassion is supposedly aroused by suffering and the murder victim would appear to be the one suffering and not the murderer, but it is theoretically possible and has even been documented as having occurred. In his testimony during his trial Adolf Eichmann expressed compassion for the Nazis who had the unfortunate duty of killing Jews and relief that he was not charged with carrying out this duty personally. This may be considered a case of possessing compassion for the wrong parties accompanied by an absence of compassion for the deserving parties. In this case, Kupperman's insightful objection is well based and one must be mindful that the recipients of one's compassion are truly deserving. From this point of view, it may well be said, compassion is not enough. One must also possess the capacity for the recognition of evil doing.
3. *Cf.*, Robert E. Allinson, 'An Overview of the Chinese Mind', Robert E. Allinson, (ed.), *Understanding The Chinese Mind: The Philosophical Roots*, New York: Oxford University Press, 2000, Tenth Impression, Chapter one, pp. 1-25.
4. Plato, *Republic*, Book II.

5. Robert E. Allinson, 'The Homogeneity and the Heterogeneity of the Concept of the Good in Plato', *Philosophical Inquiry, An International Quarterly*, Vol. IV, no. 1, Winter, 1982, pp. 30-39.
6. Plato, *Republic*, 439e-440a.
7. Kurt Baier's arguments that the questioner relies on reason to ask his question, and thus begs the question are not fully adequate. The questioner might wish to rely upon the logic of rationality in other domains, such as in the domain of asking questions and receiving answers, but may still not wish to exercise reason in the ethical domain. One remembers Dostoyevsky's underground man who delights in willing that 2 plus 2 equals 5: 'I admit that twice two makes four is an excellent thing, but if we are to give everything its due, twice two makes five is sometimes a very charming thing too.' *Notes From Underground*, IX. Most importantly, however, the attempt to give reasons for morality diminishes the ultimacy of moral truth. While there may be good reasons for being moral ('honesty is the best policy'), and morality may serve other ends, these good reasons and other ends can never supplant the moral realm of truth and as a result can never be replacement explanations for morality. For a different point of view, *Cf.*, Kurt Baier, *The Moral Point of View*, Ithaca: Cornell University Press, 1958, Chapter 7. For a view which reaches a similar conclusion but from a different process of reasoning, *Cf.*, H. A. Prichard, 'Does Moral Philosophy Rest on a Mistake?', *Mind*, Vol. 21, 1912. For Prichard, the emphasis is on the obligation rather than on the moral truth.
8. *Cf.*, Robert E. Allinson, 'The Banality of Evil and The *Challenger* Launch Decision', Letter to the Editor, *The Chronicle of Higher Education*, Vol. XLII, No. 33, April 26, 1996, pp. B3, B6. *Cf.*, Robert E. Allinson, Review Essay on the *Challenger* Disaster, *Society*, Volume 35, Number 1, pp. 99-102, Transaction Publishers, November/December, 1997. Which is worse, relativism before or after Auschwitz? Surely relativism is worse after Auschwitz. If after Auschwitz one persists not only in crediting relativism, but parading it as an ardent believer (whether or not ironically as Eichmann paraded his beliefs during the Nuremberg trials), then if a relativist is not a moral monster, the words 'moral monster' are devoid of meaning. No, it is far worse that relativism exists after Auschwitz than that it existed before. By now, one should have learned one's lesson. Auschwitz was inexcusable. If one can be a relativist now and ask for proofs of morality, then evil has been completely normalized and banalized. How can intellectuals seriously debate such a question as why be moral when Freud's four elderly sisters, all in their eighties, could be taken to Auschwitz and butchered with the rest of their fellow creatures? The question has often been raised as to why Auschwitz (and the Nazis) have been singled out for their barbarism. Needless to say, there are other examples of human atrocities such as the Japanese Rape of Nanjing of 1938-9 which included unspeakable and unthinkable acts of cruelty. The Nazi horrors have been copiously documented and serve as chilling reminders but are no means the only examples of man's cruelty to man.

Jane Goodall speaks to the issue of human cruelty in *Through a Window, thirty years with the Chimpanzees of Gomba*: '... only we, I believe, are capable of real cruelty - the deliberate infliction of physical or mental pain on living creatures despite, or even because of, our precise understanding of the suffering involved. Only we are capable of torture. Only we, surely, are capable of evil' (p. 180). One cannot assume, for example, that the cat is aware of the suffering of the mouse. The cat is not even aware of its own image in the mirror. No other animal takes positive pleasure in the exercise of cruelty upon members of its own kind. With the exception of certain rodents, no other vertebrate habitually destroys members of its own kind. Hitler is not the only example of human cruelty. According to Gibbon, Basil the Second, the conqueror of the Bulgarians, deprived fifteen thousand captives of their sight and left one of each hundred with one eye so that they could conduct the rest of their century to the king. But Hitler's immorality cannot be compared to the atrocities committed during times of war. The frightening aspect of Hitlerian atrocity was the premeditated and coldly calculated cruelty of it on such a grand scale. Imagine constructing factories the sole purpose of which was to kill people and burn them up. The Nazis were no ordinary impulse serial killers. It was the fine art of mass murder systematically and immaculately planned in advance. The mass killing of the Jewish people was not a sporadic outburst of emotional origin. It was the result of elaborately premeditated massive intelligence placed at the service of evil. Gas chambers were built by learned engineers. Children were poisoned by educated physicians. Infants were killed by trained nurses. Women and babies were shot and burned by college graduates. One and a half million Jewish children: 9 out of 10 Jewish children in Europe were murdered. Chelmno, Belzec, Sobibor and Treblinka were sites whose sole function was *immediate* extermination. Furthermore (need there be a furthermore?), it was not simply the intended extermination of everyone Jewish, and thus wiping out an entire people, as if that were not evil or heinous enough, but it was the manner of humiliation and torture that preceded and accompanied this intention. For example, the cruelty of the overcrowding of men, women and children and infants in the cattle cars of the death trains without food, water, fresh air or toilets such that twenty per cent would die in the three days (think of summer days!) the journey would require before the arrival of the train at the death camps. Or, the slow torture of starvation that the inmates of the death camps were forced to endure before their selection for the furnaces such that the soldiers of the liberating armies who arrived at the death camps were horrified at the virtual and real skeletons they met. Or, the arbitrary separation of children from parents so that mothers could be murdered in place of their children and children murdered in place of their fathers. It must be remembered that it was not simply the acts of soldiers on a battlefield or the acts of the military isolated from the watchful eyes of ordinary citizenry. It was the application of modern industrial technology to the planned extermination of a people. Think of the long-term planning:

the construction of the death camps and the killing facilities; the organization of railroads and the preparation and shipping of insecticides to exterminate the people; the years of killing. Hitler was the first member of our unhappy species to exterminate six million Jews with insect exterminators. Everything was planned down to the decimal. Even the people were all numbered (by permanent tattoos) and their numbers recorded in books. That was how it was known how many people were slaughtered. In this way, the organization of the Nazi Germans was their undoing. But the Nazi crime which Winston Churchill referred to as 'probably the greatest and most horrible crime ever committed in the history of the world' was not simply malice delivered with a high degree of organizational efficiency. Ernest A. Rappaport, in his *Anti-Judaism, A Psychohistory*, Chicago: Perspective Press, 1975, wrote that, according to Otto Fenichel, anti-Judaism is a paranoiac form of schizophrenia. The anti-Jew is in need of delusions of persecution in order to prove the over-compensated elevated position of his self. There is an excellent chapter on Hitler as possessing an obsessional paranoid delusion. Eichmann is referred to as Adolf Junior, the traveling salesman of genocide.

9. *Book of Job*, 'His wife said to him, "Are you still holding on to your integrity?" ' II, 9. While the story of Job has been construed as an object lesson in the ultimacy of faith, it is also a testament to the ultimacy of moral commitment. Job represents the ultimacy of the moral man in the immoral universe. And the morality of Job does not arise from a sense of obligation. To say that Job acted from a sense of obligation would be tantamount to saying that Einstein upon hearing the playing of Yehudi Menuhin, spoke from a sense of obligation when he said, 'now I know that G-d exists'. Einstein's feeling about Menuhin's music did not arise from a sense of duty. Nor does one need a reason to feel moved when one reads the epic poetry of Homer. Job's moral action is no less valid when it is its own reason for being.

Moderato
Expressively, but not dragging

3 The Metaphysical Foundation of the Phenomenological Structure of Ethics: Ethics as Regional Metaphysics

Metaphysics and Ethics

In discussing ethics, it may have come as a surprise that much of the discussion focused on questions surrounding the issue of human nature rather than ethics as such. However, the objective of raising such questions was to shed some light on problems in ethics rather than the theoretical objective of arriving at some knowledge concerning human nature.

Ethics as Regional Metaphysics

Of course, in one sense, ethics as constituting part of metaphysics is not a new idea if one considers Spinoza, whose major metaphysical treatise is at once an ethics and is titled as such. One difference, of course, is that if one carries out regional metaphysics, one can consider ethical truths apart from knowing the nature of metaphysical truths from other domains. For Spinoza, all things follow from the infinite nature of G-d with necessity. For Spinoza, there is no freedom in nature and all the causes in nature are traceable to the necessity of the Divine nature. In regional metaphysics, there is no need to trace the metaphysical truths in ethics back to the necessity of the Divine nature. It appears that regional domains may be treated as consisting of metaphysical truths without having to posit the existence of interrelationships of a deductive nature between or among some or all metaphysical regions. The truths that obtain within regions can be known as truths apart from the consideration of how they may be connected to the truths of other regional domains. What is most important is that the nature of metaphysical truth within one region may not be identical with the nature of metaphysical truth within another region.

Metaphysical Truths in Ethics

All moral harm-doing as performed by human beings is the result either of the excess of healthy human appetites, or the predominance of negative human emotions, nearly any of which can be traceable back to their origin in fear, hurt, or the desire to protect. Whether or not in the end one subscribes to this position, or to some modified form of it, it is of interest to recognize this to be a metaphysical position, and thus as constituting a region of metaphysical concern. How may it be recognized as a metaphysical position? First of all, it is stated to be true of the whole of its region. Secondly, while it is not always put forward as such, one may take it in its most meaningful form to be intended as a necessary truth. Thirdly, it is impossible to prove by ordinary empirical means and hence is scientifically non-verifiable. Fourthly, while a philosopher may first note it to be true, it is phenomenologically testable in the crucible of every person's experience. While it could be possible to characterize metaphysical truths of this region as synthetic *a priori* truths, such a characterization may prove unenlightening at this point. It is best to point out that such truths are known as true through deep and practised reflection on experience, but they are by no means empirical generalizations. Metaphysical truths of this region may be said to be known by a philosopher in an analogous fashion to how an experienced psychoanalyst may know truths within her or his region. Insofar as the philosopher is a physician of the soul, the philosopher no less than the psychoanalyst may be said to possess clinical insight. Fifthly, and this is true of this metaphysical region, although it may not be true of all other metaphysical regions, the metaphysical truth which belongs to this region is one which is at once prescriptive and descriptive.

It may sound odd that a truth can be both prescriptive and descriptive, but in the realm of metaphysical truths that belong to the ethical region, such is the status of metaphysical truth. The joint status of metaphysical truth as being both prescriptive and descriptive arises because the metaphysical truth of this region is descriptive of human beings who possess the power to bring descriptions into actualization. The prescriptive/descriptive categorization is neither intended to take into account what other metaphysicians have done in the past, or must do, nor is it taken to have to be true of all metaphysical domains; it is a categorization that describes the ethical region and describes the ethical region in intent, at least, only within this present metaphysical approach. Whether it describes other normative systems of ethics is another matter entirely.

Fact and Value

Ever since Hume, it has become commonplace that no 'ought' can be deduced from an 'is' and philosophers have become very wary of expropriating statements from the realm of fact and applying them to the realm of value. However, in the case of ethics as defined above, metaphysical statements are at once descriptive (factually grounded) and prescriptive (fall into the class of statements in which one will inevitably discover either the word 'should' or the word 'ought'). In the special case and style of ethics as has been advanced above, one does not attempt to borrow truths from the realm of fact and then apply them to the realm of value, but rather one attempts to demonstrate that some congruence exists between certain (not all) facts and certain values. An example of the argument from congruence may be seen if one analyzes the claim that human nature is fundamentally good.

When such a claim is made, the first test to be applied is the phenomenological test. One first of all asks oneself, in the manner of Anne Frank, is human nature fundamentally good? If the results of such self-questioning are inconclusive, one may attempt to define more clearly what is or should be meant by 'good' and what by 'evil'. Evil actions might be considered to be that class of actions that are hurtful to oneself (out of respect for no moral principle), to one's fellow human beings, to dumb beasts, and so on (the categories may be altered to suit changing circumstances such as the discovery of intelligent life on other planets). In such a definition of evil, one may take intentional harming of oneself or one's fellow human beings (this may stand for all relevant creatures) to be a completely replaceable (not substitutable) expression for 'evil', such that the word 'evil' may be dropped from the vocabulary.

The question can now once more be raised, do human beings ever mean to intentionally hurt others or themselves, and if so, under what special conditions? Such a question may then generate various answers. One may intentionally hurt others when they are threatening one's livelihood, or those of one's family, or one's nation, religious group, and so on. One may hurt oneself when suffering from psychological distress. If cases of psychological deviance are excluded, the question can then be amended to read, do human beings intentionally do gratuitous harm to themselves or others? Now, apart from those cases which obviously are a result of special pathological proclivities on the part of certain disturbed individuals, and whole cultures in which a mass of individuals act in concert or compliance with a single individual or a small group of individuals, and such cases of self-harm which fall under the bio-psychological deduction of the emotions discussed above, the general introspective answer will more clearly become 'no'.

After it has been determined by this phenomenological self-test, that non-pathological human beings do not essentially intend gratuitous harm-doing to themselves or to fellow humans, one may then ask, whence does such a value or attitude as the prohibition against gratuitous harm-doing arise? Is it inborn? Is it a product of education? After extended reflection and discussion, one may reach the conclusion that the prohibition against gratuitous harm-doing is a result of education and culture and/or arises from a nature that is inclined against gratuitous harm-doing and is furthered in its inclination against harm-doing by having learned values apposite to its original inclination. If, in either case, one further considers that the prohibition against harm-doing is one which is a cultural universal, one will have succeeded in discovering an ethical value that is considered to be true both in the sense that it is part of human nature and considered right that it should form a part of human nature.

But, the question remains, how can one know that such a value as the prohibition against harm-doing both is a part of human nature and considered right or good that it should form such a part? If the prohibition against gratuitous harm doing is a fundamental part of human nature in some respect and is enhanced by teaching, then it is possible to reach the conclusion that the prohibition against gratuitous harm-doing is a metaphysical truth. Or, it can be argued that whether or not it is a metaphysical truth, it is better to think that it is, and that for the sake of world peace, humankind will be all better off if such a prohibition is held to be a norm of human action. These approaches, however, will not find much favor with those who consider that human beings are fundamentally evil, and that nothing can be done to alter this disposition.

When one who argues for the fundamental goodness of human nature encounters one who argues the opposite, it can be said that the interlocutors have reached a metaphysical impasse. There are those who believe that human beings are fundamentally evil, and thus need to have ethical systems devised that address this proclivity, and there are those who believe that human beings are fundamentally good and require the sort of systems or approaches that take this into account. For the time being, those who believe in neutrality or in some combination of both good and evil may be omitted in order to produce a greater simplicity of argument.

Suppose one entertains the belief system that human beings are evil. Even if one takes up the second question, i.e., from whence do one's fundamental belief systems arise, some might argue that even if one's belief that all human beings were evil were a belief inherited from an education that human nature was evil, this was indeed in accord with the factual truth. Thus, the search into possible causes of belief systems still leaves the question unanswered as to which belief system is the true one. This is the point at which the factual question turns into the evaluative one. Because the metaphysical approach introduced in this present work subscribes to the belief that it is better to consider human nature to be good rather than its opposite, it can be argued that the claim that human nature is good is not only a descriptive claim, but a prescriptive claim that requires consent and active support to complete its truth value. While this may at first appear to violate the demands of metaphysics to consider only objective fact, it must be remembered that in the regional domain of ethics, the subject matter of inquiry and the inquiring subject are one and the same. The subject inquirer, thus, possesses the right, nay, the obligation to project a good direction for the subject matter.

In the regional metaphysical domain of ethics, the subject knower possesses the power and the responsibility to determine at least partially the nature of the subject matter. It is at this juncture that the domains of fact and value intersect and questions of fact and questions of value properly overlap and properly interrelate. That human beings are all good is a fact that can follow upon (or be produced by) the universal adoption of such a value (and whatever practices must also accompany the adoption of such a value), just as much as the value that human beings are good can be founded upon the fact of discovered human goodness. The origin of the goodness does not greatly matter. In the case of the subject of human nature, a clear-cut distinction between facts and values cannot be drawn, so that the very question as to how the value of goodness arises is not capable of being answered in a fully definitive sense, nor should it be so capable of being answered. If it could be said that goodness were a value, and not a fact, or vice-versa, the claim would be set forth that metaphysics was either prescriptive or descriptive. On the contrary, in the case of the region of ethics, metaphysics is a blend of both the descriptive and the normative. One's very ideas about human nature partially determine that nature, and human nature is partially determinative of one's ideas.

Human beings are at least in part, self-constituted entities, and these self-constituted entities are composed both of a factual and an evaluative component. How in particular the direction of human nature can be argued to be good, rather than evil, is dependent upon the right understanding of the bio-psychological derivation of the emotions. Of immediate interest is the discovery that the position that 'human nature is good' is a metaphysical position, and that when such a claim is made, what is meant is that this is a truth that is true of human beings as human beings, and therefore true of all human beings as such. This truth claim is both prescriptively true and consequently (as self-determined) descriptively true, and therefore must also have been descriptively true as a possibility that becomes true whenever it is prescriptively evoked. Whether or not one wishes to subscribe to this metaphysical claim, it remains true that such a claim that appears most often as a part of a philosopher's ethical or political system is indubitably a metaphysical one. Whether it may be shown to be a non-metaphysical claim by a further inquiry or by more refined empirical science is another issue altogether. At this stage of knowledge, it exists as a metaphysical truth claim belonging to the ethical region of metaphysical domains.

A Note on the Bio-psychological Deduction of the Emotions

If the deduction that appeared in the Theoretical Foundations of Ethics is correct, then it may be said that metaphysical statements that belong to the region of ethics are no longer moot statements. Within the rubric adopted above of the descriptive/prescriptive blend of truth, certain metaphysical statements can be considered to be true and other metaphysical statements can be considered to be false. The statement, 'human nature is good' can be considered to be true, and the statement, 'human nature is evil' can be considered to be false. The values of truth or falsity can be assigned correctly in this case by virtue of an understanding of the bio-psychological deduction and an understanding of the importance of the prescriptive dimension. It is not the case then, that metaphysical statements are so broad or so technically couched so as to be without truth-value or means of justification. The case of the statement 'human nature is good' appears to be an example of a metaphysical truth which both can be discovered by phenomenological introspection, and for which an argument can be provided to account for its all too frequent and in most cases abysmal lack of empirical instantiation.

There need not be a conflict between the findings of phenomenology, and a metaphysical inquiry into the origin of values. In fact, it would appear that there should be congruence rather than conflict, for a good metaphysics should offer an explanation for why it is possible to have the introspections that one possesses, and by the same token, one's introspections should be capable of finding their justification in metaphysical inquiry. Phenomenological introspection is of course an independent realm of inquiry and truth finding and does not need to defer to metaphysics in order to validate its own found data. But it is interesting to note that phenomenology need not stand isolated and bereft of support from other disciplines of knowledge. In fact, as should be somewhat evident in the bio-psychological deduction, phenomenological truths and metaphysical deductions stand side by side as links within an argument that is both phenomenological and metaphysical. The demands of phenomenology and rationality coalesce in their findings, which is as it should be in any metaphysics worthy of the name.

Da Capo

Bibliography

Alexander, H. G., (ed.), *The Leibniz-Clarke Correspondence*, New York: Philosophical Library, 1956.

Alexander, S., *Space, Time and Deity*, Volumes I and II, London: Macmillan and Co., Ltd., 1934.

Allinson, Robert E., *A Metaphysics for the Future*, Aldershot: Ashgate, 2001.

Allinson, Robert E., 'The Banality of Evil and The *Challenger* Launch Decision', Letter to the Editor, *The Chronicle of Higher Education*, Vol. XLII, No. 33, April 26, 1996, pp. B3, B6.

Allinson, Robert E., 'The Confucian Golden Rule: A Negative Formulation', *Journal of Chinese Philosophy*, Vol. 12, No. 3, September 1985.

Allinson, Robert E., 'The Homogeneity and the Heterogeneity of the Concept of the Good in Plato', *Philosophical Inquiry, An International Quarterly*, Vol. IV, No. 1, Winter, 1982.

Allinson, Robert E., (ed.), *Understanding the Chinese Mind: The Philosophical Roots*, New York: Oxford University Press, 1989, 2000, Tenth Impression.

Allison, Henry E., *The Kant-Eberhard Controversy*, Baltimore and London: The Johns Hopkins University Press, 1973.

Baier, Kurt, *The Moral Point of View*, Ithaca: Cornell University Press, 1958.

Barker, Jerome H., Cosmides, Leda and Tooby, John, (eds.), *The Adapted Mind, Evolutionary Psychology and the Generation of Culture*, New York and Oxford: Oxford University Press, 1992.

Bary, Wm. de, Theodore, (ed.), *Sources of Chinese Tradition*, New York: Columbia University Press, 1960.

Berlin, Isaiah, *Russian Thinkers*, Henry Hardy and Eileen Kelly, (eds.), Harmondsworth: Penguin Books, 1979.

Bohr, Niels, *Essays 1932-1957 on Atomic Physics and Human Knowledge, The Philosophical Writings of Niels Bohr*, Vol. II, Woodbridge, Connecticut: Ox Bow Press, 1987.

Boslough, John, *Beyond the Black Hole, Stephen Hawking's Universe*, UK: Collins, 1989.

Capra, Fritjof, *The Tao of Physics*, Boston: Shambhala, 1991.

Collingwood, R. G., *The New Leviathan*, London: Oxford University Press, 1958.

Cornford, F. M., *Plato and Parmenides*, London: Routledge & Kegan Paul, 1939.

Cua, A. S., 'The conceptual aspect of Hsün Tzu's philosophy of human nature', *Philosophy East and West*, 27, No. 4, October 1977.

Cua, A. S., 'The quasi-empirical aspect of Hsün-Tzu's philosophy of human nature', *Philosophy East and West*, 28, No. 1, January 1978.

Dawkins, Richard, *The Selfish Gene* (new edition), Oxford: Oxford University Press, 1989.

Deutsch, David, 'Quantum Mechanics Near Closed Timelike Lines', *Physical Review* D, Vol. 44, No. 10, November 15, 1991.

Deutsch, David, and Lockwood, Michael, 'The Quantum Physics of Time Travel', *Scientific American*, March 1994.

Dicker, Georges, *Descartes, An Analytical and Historical Introduction*, New York: Oxford University Press, 1993.

Dummett, Michael, *The Logical Basis of Metaphysics*, Cambridge, Massachusetts: Harvard University Press, 1991.

Doney, Willis, (ed.), *Descartes, A Collection of Critical Essays*, Garden City, New York: Doubleday & Company, 1967.

Doney, Willis, (ed.), *Eternal Truths and the Cartesian Circle, A Collection of Studies*, New York and London: Garland Publishing, Inc., 1987.

Dostoyevsky, Fyodor, *Notes from Underground, Poor People, The Friend of the Family*, Constance Garnett, (trans.), New York: Dell Publishing Co., 1960.

Durden-Smith, Jo, and Simone, Diane de, *Sex and the Brain*, London: Pan Diane de Books, 1983.

Erikson, Erik, *Gandhi's Truth on the Origins of Militant Nonviolence*, New York: Norton, 1969.

Feyerabend, Paul K., *Science in a Free Society*, London: New Left Books, 1979.

Fink, Eugen, and Husserl, Edmund, *Sixth Cartesian Meditation: the idea of a transcendental theory of method*, with textual notations by Edmund Husserl, Ronald Bruzina, (trans.), Bloomington: Indiana University Press, 1995.

Flood, R. and Lockwood, M., (eds.), *The Nature of Time*, Basil Blackwell, 1986.

Fodor, Jerry, A., and LePore, Ernest, (eds.), *Holism: A Consumer Update*, Amsterdam: *Grazer Philosophische Studien*, 1993.

Folse, Henry, J., *The Philosophy of Niels Bohr, The Framework of Complementarity*, Amsterdam: Elsevier Science Publishers, 1985.

Frankl, Viktor, E., *Man's Search For Meaning: An Introduction to Logotherapy*, New York: Pocket Books, 1972.

Frege, Gottlob, *The Foundations of Arithmetic*, J. L. Austin, (trans.), Evanston: Northwestern University Press, 1980.

Fromm, Erich, *The Anatomy of Human Destructiveness*, New York: Holt, Rinehart and Winston, 1973.

Gardner, Howard, *Frames of Mind, The Theory of Multiple Intelligences*, New York: Basic Books, 1983.

Garntsev, Mikhail, 'The Problem of Absolute Freedom in Descartes', *Russian Studies in Philosophy*, Vol. 37, No. 2, Fall, 1998.

Goodall, Jane, *Through a window, Thirty Years with the Chimpanzees of Gombe*, London: Penguin Books, 1990.

Gram, Moltke, S., *Kant: Disputed Questions*, Chicago: Quadrangle Books, 1967.

Gribben, John, *In Search of Shröedinger's Cat, Quantum Physics and Reality*, London: Corgi Books, 1984.

Gribben, John, *In Search of the Edge of Time*, London: Black Swan Books, 1992.

Haldane, Elizabeth S. and Ross G. R. T., *The Philosophical Works of Descartes*, Vol. 1, Cambridge: Cambridge University Press, 1967.

Handyside, John (trans.), *Kant's Inaugural Dissertation and Early Writings on Space*,Westport: Hyperion Press, 1988.

Hardy, Henry and Kelly, Aileen, (eds.), *Russian Thinkers*, Harmondsworth: Penguin Books, 1979.

Hartshorne, Charles, *Creative Synthesis and Philosophical Method*, La Salle, IL: Open Court, 1970.

Harris, Errol E., *Fundamentals of Philosophy, A Study of Classical Texts*, New York, Holt Rinehart and Winston, 1969.

Hawking, Stephen, *A Brief History of Time*, New York: Bantam Books, 1989.

Hawking, Stephen, *Black Holes and Baby Universes and Other Essays*, New York: Bantam Books, 1994.

Hawking, Stephen, and Penrose, Roger, *The Nature of Space and Time*, Princeton: Princeton University Press, 1996.

Hawking, Stephen and Penrose, Roger, 'The Nature of Space and Time', *Scientific American*, July 1996.

Hegel, G. W. F., *The Encyclopaedia of the Philosophical Sciences*, William Wallace, (trans.), Oxford: The Clarendon Press, 1892.

Hegel, G. W. F., *Phänomenologie des Geistes*, Hamburg: Felix Meiner Verlag, 1952.

Hegel, G. W. F., *The Phenomenology of Mind*, J. B. Baillie, (trans.), New York and Evanston: The Academy Library, 1967.

Heisenberg, Werner, *Physics and Beyond, Encounters and Conversations*, London: George Allen & Unwin, Ltd., 1971.
Hillman, James, *The Soul's Code, In Search of Character and Calling*, New York: Random House, 1996.
Hintikka, Jaakko, 'Cogito, Ergo Sum: Inference or Performance?', *The Philosophical Review*, Vol. 71, 1961, pp. 3-33.
Höffding, Harald, *The Problems of Philosophy*, G. M. Fisher, (trans.), New York: Macmillan, 1905.
Hollis, Martin and Nell, Edward, J., *Rational economic man, a philosophical critique of neo-classical economics*, Cambridge: Cambridge University Press, 1975.
Howard, Don, *Complementarity and Ontology*, Doctoral Dissertation, Ann Arbor: University Microfilms, 1979.
Hsün-Tzu, *The Works of Hsuntze*, Homer H. Dubs, (trans.), Confucius Publishing Co., 1972.
Huck, Geoffrey J. and Goldsmith, John A., *Ideology and Linguistic Theory, Noam Chomsky and the Deep Structure Debates*, London and New York: Routledge, 1995.
Hui Neng, *Sutra Spoken by the Sixth Patriarch On the High Seat of "The Treasure of the Law*", Wong Mou-Lam, (trans.), Hong Kong: Hong Kong Buddhist Book Distributor, 1953.
Husserl, Edmund, *Ideas, General Introduction to Pure Phenomenology*, W. R. Boyce Gibson, (trans.), New York: Macmillan Publishing Co., Inc., 1962.
Husserl, Edmund, *The Paris Lectures*, Peter Koestenbaum, (trans.), The Hague: Martinus Nijhoff, 1970.
Jammer, Max, *The Conceptual Development of Quantum Mechanics*, New York: McGraw-Hill Book Company, 1966.
Jones, W. T., *A History of Western Philosophy*, New York: Harcourt, Brace and Company, 1952.
Kant, Immanuel, *Kritik der reinen Vernunft*, Hamburg: Felix Meiner Verlag, 1956.
Kant, Immanuel, *Critique of Pure Reason*, Norman Kemp Smith, (trans.), New York: St. Martin's Press, 1965.
Kant, Immanuel, *Philosophical Correspondence 1759-99,* Arnulf Zweig, (ed. and trans.), Chicago: The University of Chicago Press, 1967.
Kant, Immanuel, *Anthropology from a Pragmatic Point of View* (1798), Mary J. Gregor (trans.), The Hague: Nijhoff, 1974.
Kant, Immanuel, *Kant's Inaugural Dissertation and Early Writings on Space*, John Handyside, (trans.), Westport: Hyperion Press, 1988.

Köhler, Wolfgang, Ella Winter, (trans.), *The Mentality of Apes*, New York: Harcourt, Brace and Company, 1931.

Kragh, Helge, *Cosmology and Controversy, The Historical Development of Two Theories of the Universe*, Princeton: Princeton University Press, 1996.

Kragh, Helge, *Quantum Generations, A History of Physics in the Twentieth Century*, Princeton: Princeton University Press, 1999.

Kramer, Elmar J., (ed.), *Interpreting Arnauld*, Toronto: University of Toronto Press, 1966.

Küng, Hans, *Judaism*, Munich: Piper Verlag, 1991.

Kupperman, Joel, J., *The Foundations of Morality*, London: Allen & Unwin, 1983.

Latour, Bruno and Steve Woolgar, *Laboratory Life*, Beverly Hills: Sage Publications, 1979.

Latour, Bruno, *Science in Action*, Milton Keynes: Open University Press, 1987.

Lawson, Hilary and Appignanesi, (eds.), *Dismantling Truth, Reality in the Post-Modern World*, London: Weidenfeld and Nicolson, 1989.

Lewis, David, 'The Paradoxes of Time Travel', *American Philosophical Quarterly*, April, 1976, Vol. 13, No. 2.

Lineham, Marsha M., *Understanding Borderline Personality Disorder*, New York and London: The Guilford Press, 1995.

Loeb, Lewis, E., 'The Cartesian Circle', in John Cottingham, (ed.), *The Cambridge Companion to Descartes*, Cambridge: Cambridge University Press, 1992.

Lorenzen, Paul, *Normative Logic and Ethics*, Mannheim/Zurich, *Bibliographisches Institut*, 1969.

Mates, Benson, *The Philosophy of Leibniz, Metaphysics and Language*, New York: Oxford University Press, 1986.

McKeon, Richard, (ed.), *The Basic Works of Aristotle*, New York: Random House, 1941.

Midgley, Mary, *Beast and Man*, Sussex: Harvester Press, 1978.

Miller, Alice, *For Your Own Good, Hidden Cruelty in Child-Rearing and the Roots of Violence*, Hildegarde and Hunter Hannum, (trans.), New York: Farrar, Strauss and Giroux, 1985.

Montagu, Ashley, *The Nature of Human Aggression*, New York: Oxford University Press, 1976.

Moore, Charles A. (ed.), *The Chinese Mind, Essentials of Chinese Philosophy and Culture*, Honolulu: East-West Center Press, University of Hawaii Press, 1967.

Moser, Paul K., (ed.), *A Priori Knowledge*, Oxford: Oxford University Press, 1987.

Naess, Arne, and Hannay, Alistair, (eds.), *Invitation to Chinese Philosophy*, Oslo: Scandinavian University Books, 1972.

Ortony, Andrew, Clore, Gerald, L., and Collins, Allan, *The Cognitive Structure of Emotions*, Cambridge, Cambridge University Press, 1988.

Pais, Abraham, *Niels Bohr's Times, In Physics, Philosophy, and Polity*, Oxford: Clarendon Press, 1991.

Penrose, Roger, *Shadows of the Mind*, New York: Oxford University Press, 1994.

Penrose, Roger, *The Emperor's New Mind*, New York: Oxford University Press, 1989, 1990.

Penrose, Roger, Shimony, Abner, Cartwright, Nancy and Hawking, Stephen, The *Large, The Small and The Human Mind*, Cambridge: Cambridge University Press, 1999.

Petruccioli, Sandro, *Atoms, Metaphors and Paradoxes, Niels Bohr and the construction of a new physics*, Ian McGilvray, (trans.), Cambridge: Cambridge University Press, 1993.

Plutchik, R., *The emotions: Facts, theories, and a new model*, New York: Random House, 1962.

Popkin, Richard, *The History of Scepticism from Erasmus to Descartes*, New York: Harper and Row, 1964.

Rappaport, Ernest A., *Anti-Judaism, A Psychohistory*, Chicago: Perspective Press, 1975.

Reid, William Lewis, III, *An Examination of Descartes' Evil Demon Argument*, Doctoral Dissertation, Ann Arbor: University Microfilms, 1979.

Rorty, Richard, *Essays On Heidegger And Others, Philosophical Papers*, Volume 2, New York: Cambridge University Press, 1991.

Rosenbaum, Ron, *Explaining Hitler, The Search For The Origins of His Evil*, New York: Random House, 1998.

Ruse, Michael, 'Morality of the Gene', *The Monist*, Vol. 67: 2, April 1984.

Russell, Bertrand, *The Problems of Philosophy*, Oxford: Oxford University Press, Seventeenth Impression, 1943.

Schilpp, Paul Arthur, (ed.), *Albert Einstein, Philosopher-Scientist*, Library of Living Philosophers, Vol. VII, La Salle, Illinois: Fifth Printing, 1991.

Searle, John, R., *Minds, Brains and Science*, Cambridge: Harvard University Press, 1985.

Seebohm, Thomas M., and Kockelmans, Joseph J., (eds.), *Kant and Phenomenology*, Washington, D. C.: Center for Advanced Research in Phenomenology and University Press of America, 1984.

Singer, Peter, *The Expanding Circle, Ethics and Sociobiology*, Oxford: Oxford University Press, 1983.
Sklar, Lawrence, *Philosophy of Physics*, New York: Oxford University Press, 1992.
Slavinska-Holy, Nonna, *Borderline and Narcissistic Patients in Therapy*, Madison: International Universities Press, Inc., 1988.
Spinoza, Benedict de, *Ethics*, James Gutmann, (ed.), New York: Hafner Publishing Co., 1963.
Todhunter, Isaac, (ed.), *Euclid's Elements*, London and New York: Everyman's Library, 1967.
Waal, Frans De, *Good natured, The Origins of Right and Wrong in Humans and Other Animals*, Cambridge: Harvard University Press, 1996.
Wheatley, Margaret J., *Learning About Organizations from an Orderly Universe*, San Francisco: Berrett-Koehler Publishers, 1992.
Williams, Bernard, *Descartes, The Project of Pure Enquiry*, Harmondsworth and New York: Penguin Books, 1986.
Wolfson, Harry Austryn, *The Philosophy of Spinoza*, Volumes I and II, Cambridge: Harvard University Press, 1934.
Wolin, Richard, *Walter Benjamin, An Aesthetic of Redemption*, New York: Columbia University Press, 1982.

Index

For Product Safety Concerns and Information please contact our EU representative GPSR@taylorandfrancis.com Taylor & Francis Verlag GmbH, Kaufingerstraße 24, 80331 München, Germany

Batch number: 10725802

Printed by Printforce, the Netherlands